CAMBRIDGE LIBRARY COLLECTION

Books of enduring scholarly value

Technology

The focus of this series is engineering, broadly construed. It covers technological innovation from a range of periods and cultures, but centres on the technological achievements of the industrial era in the West, particularly in the nineteenth century, as understood by their contemporaries. Infrastructure is one major focus, covering the building of railways and canals, bridges and tunnels, land drainage, the laying of submarine cables, and the construction of docks and lighthouses. Other key topics include developments in industrial and manufacturing fields such as mining technology, the production of iron and steel, the use of steam power, and chemical processes such as photography and textile dyes.

A Treatise on Engineering Field-Work

In the 1840s, the civil engineer Peter Bruff (1812–1900) designed what was then the largest brick structure in Britain, the 1,000-foot-long Chappel Viaduct in Essex. He went on to become a railway entrepreneur and developer, and was responsible for the creation of the resort town Clacton-on-Sea, where he also designed many of the buildings. In this illustrated guide, first published in 1838 and here reissued in the revised and expanded two-volume second edition of 1840–2, he discusses the theory and practice of surveying (calculating the accurate position of points in the landscape) and levelling (calculating the accurate height of points). Volume 2 covers levelling; Bruff gives a brief overview of the theory, then describes the typical equipment used. By discussing such examples as the levelling of a projected railway route, he explains what information should be recorded, and how to avoid common errors of technique that affect accuracy.

Cambridge University Press has long been a pioneer in the reissuing of out-of-print titles from its own backlist, producing digital reprints of books that are still sought after by scholars and students but could not be reprinted economically using traditional technology. The Cambridge Library Collection extends this activity to a wider range of books which are still of importance to researchers and professionals, either for the source material they contain, or as landmarks in the history of their academic discipline.

Drawing from the world-renowned collections in the Cambridge University Library and other partner libraries, and guided by the advice of experts in each subject area, Cambridge University Press is using state-of-the-art scanning machines in its own Printing House to capture the content of each book selected for inclusion. The files are processed to give a consistently clear, crisp image, and the books finished to the high quality standard for which the Press is recognised around the world. The latest print-on-demand technology ensures that the books will remain available indefinitely, and that orders for single or multiple copies can quickly be supplied.

The Cambridge Library Collection brings back to life books of enduring scholarly value (including out-of-copyright works originally issued by other publishers) across a wide range of disciplines in the humanities and social sciences and in science and technology.

A Treatise on Engineering Field-Work

Comprising the Practice of Surveying,
Levelling, Laying Out Works,
and Other Field Operations
Connected with Engineering

VOLUME 2

PETER BRUFF

CAMBRIDGE
UNIVERSITY PRESS

CAMBRIDGE
UNIVERSITY PRESS

University Printing House, Cambridge, CB2 8BS, United Kingdom

Cambridge University Press is part of the University of Cambridge.
It furthers the University's mission by disseminating knowledge in the pursuit of education, learning and research at the highest international levels of excellence.

www.cambridge.org
Information on this title: www.cambridge.org/9781108071543

This edition first published 1842
This digitally printed version 2014

ISBN 978-1-108-07154-3 Paperback

The original edition of this book contains a number of colour plates, which have been reproduced in black and white. Colour versions of these images can be found online at www.cambridge.org/9781108071543

A TREATISE

ON

ENGINEERING FIELD-WORK,

COMPRISING

THE PRACTICE OF

SURVEYING, LEVELLING, LAYING OUT WORKS,

AND OTHER

FIELD OPERATIONS CONNECTED WITH ENGINEERING.

With numerous Diagrams and Plates.

BY PETER BRUFF, C.E.,

ASSOCIATE INST. CIVIL ENGINEERS.

LEVELLING.

LONDON:

SIMPKIN, MARSHALL, AND CO., STATIONERS' HALL COURT;

HEBERT, CHEAPSIDE; TAYLOR, WELLINGTON STREET, STRAND;

WEALE, ARCHITECTURAL LIBRARY, HIGH HOLBORN;

AND WILLIAMS, GREAT RUSSELL STREET, BLOOMSBURY.

1842.

LONDON:
THOMS, PRINTER, WARWICK SQUARE.

CONTENTS.

CHAPTER I.

CHAPTER II.

CHAPTER III.

CHAPTER IV.

CHAPTER V.

CHAPTER VI.

NOTICE.

I much regret the delay which has occurred in the appearance of the present Volume, but it has been unavoidable. Domestic afflictions have contributed considerably thereto; but still more has the unceasing attention requisite to the successful prosecution of many varied undertakings which I have been engaged in since the appearance of the first volume. I have far advanced with the concluding part, and trust the time which has elapsed, instead of being found injurious to its contents, will be deemed favourable, as affording me more time for consideration of the matters therein advanced, and opportunity of more fully developing my ideas.

P. B.

Charlotte Street, Bloomsbury,
June, 1842.

THE

THEORY AND PRACTICE

OF

LEVELLING.

CHAPTER I.

THEORY OF LEVELLING, GRAVITY AND THE PLUMB-LINE.—FIGURE OF THE EARTH.—CURVATURE, WITH VARIOUS FORMULÆ FOR COMPUTING.—EXAMPLES OF CORRECTIONS FOR CURVATURE.—ATMOSPHERIC REFRACTION, EFFECTS OF.—GENERAL REMARKS.

IN this part of our work we shall strictly confine ourselves to introductory matter and preliminary observations in the first chapter, in order to put the subject in a clear light before such of our readers as are ignorant of its principles, or have never given it due consideration. The figure of the earth may be understood or defined by a surface at every point perpendicular to the direction of gravity, or of the plumb-line when it is unaffected by surrounding objects. This surface is of the same form that the sea would assume if continued all round the earth, and unaffected by wind or tide; the surface of every fluid, when at rest, being perpendicular to the direction of gravity. The visible horizon of an observer on the surface of the earth is a tangent plane at the point of observation to the curve surface formed by the earth's exterior ; or in other words, the line of sight is at right angles to the direction of gravity at that point, or of the semi-diameter of the earth.

The art of levelling consists in finding or tracing a line on a given portion of the earth's surface parallel to the horizon at *all* its points,—consequently parallel to the earth's mean surface ; and any number of such points are on a true level when equidistant from the centre of the earth, considering it as a perfect sphere. Such a line would therefore be a curve; and if we were to trace curve lines, by levelling from a given point round the earth in every direction, till they returned into themselves, the superficies in which all these lines would lie, is that which we consider as the superficies of the earth. The figure which is bounded by this superficies, is that which is really measured by the combined method of astronomy and practical geometry, and is to be carefully distinguished from the *actual* figure of the earth, including all its inequalities.* But the line of sight given by the operation of levelling is similar to that of an observer,—viz. a tangent to the earth's surface at the point of observation. In the accompanying diagram B D E, &c. represents the line of sight, and B C F G, a portion of the earth's surface, its centre being at A. The line of sight being a tangent to the curved surface of the earth, is perpendicular to its semi-diameter at the point of contact B, rising always higher above the curved surface or the true line of level, the further the distance is extended; this is called the *apparent* line of level. Thus, C D is the height of the ap-

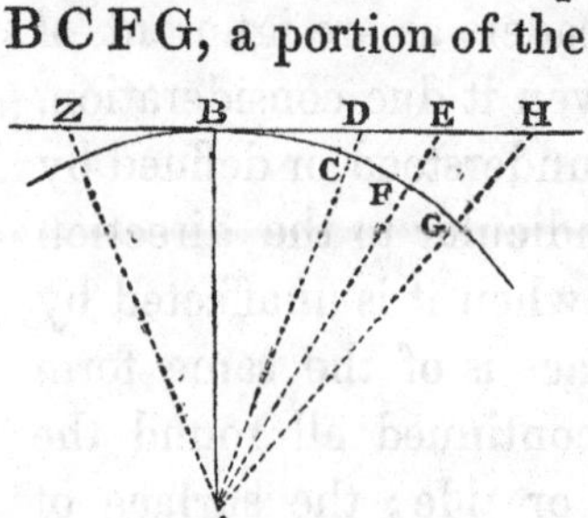

* Playfair's Natural Philosophy.

parent above the true level at the distance B C; E F, is the excess of height at the distance B F; G H, at B G, &c., the difference, it is evident, being always equal to the excess of the secant of the arc of distance, above the radius of the earth. Where the line of sight does not extend for any considerable distance, the surface may be considered as a plane, but this must by no means be the case where the points of observation are far apart. The *effect* of the curvature of the earth is to *depress* the apparent place of an object, and the measure of its quantity may be deduced from the following simple proportion, (*see last diagram*); $(2\,AC + CD) : BD :: BD :$ but the diameter of the earth (2 A C) being so great with respect to C D, at all distances to which ordinary operations of levelling extend, that 2 A C may be taken in this proportion for 2 A C + C D without sensible error; the proportion will therefore stand thus: $2\,AC : BD :: BD : CD$, whence $CD = \frac{BD^2}{2\,AC}$ or $\frac{BC^2}{2\,AC}$ nearly; or, in other words, the difference between the apparent and true level is equal to the square of the distance between the places, divided by the diameter of the earth, and consequently proportional to the square of the distance. Thus, the mean diameter of the earth being taken at 7,916 miles,* if we first take $BC = 1$ mile, then the excess $\frac{BC^2}{2\,AC}$ becomes $\frac{1}{7916}$ths of a mile, equal to 8.004 inches, or .667 of a foot, for the height of the apparent above the true level at the distance of one mile. The height of the apparent above the true level, it will be observed, increases as the square of the distance; that is to say, at two miles the difference is four times as great as it is at one mile; at three miles it is nine times as great, and so on. The following example will

* Equatorial diameter 7,924 miles; Polar diameter 7,908 miles; *mean* diameter 7,916 miles.

perhaps set this matter more clear before some of our readers :—A spirit level (see description of that instrument) is planted on a hill, and when accurately levelled, the horizontal wire of the *diaphragm* in the telescope is observed exactly to coincide with the summit of a church steeple, distant five miles ;—required the difference of level (if any) between the summit of the steeple and the ground where the level was planted, the telescope of the instrument being elevated 4.50 feet above the surface :—

Amount of depression due to curvature for 1 mile ..	.667 Feet
Square of distance ; 5 miles	25
	3335
	1334
Amount of depression due to curvature for 5 miles..	16.675 Feet
Height of instrument above the surface of ground to be added..................................	4.500
Difference of level	21.175 Feet.

From this example it will be seen, that although the apparent difference of level was only 4.50 feet (the height of the instrument), yet the summit of the steeple was found to be 21.175 feet higher than the ground where the spirit level was planted, but on account of the curvature of the earth it was apparently depressed to the same level as the centre of the telescope.

The correction for curvature may also be computed by the well-known proposition, "That in any right angled triangle, the square of the hypothenuse is equal to the sum of the squares of the other two sides ;" therefore, by taking the *sum* of the squares of the tangent line or distance, and of the semi-diameter of the earth, and extracting the square root, we obtain the hypothenuse, from which deducting the semi-diameter or radius of the earth, the height of the apparent above the true level is obtained as before.

There are several other convenient forms by which the correction for curvature may be ascertained. The following formula will be easily remembered.—Divide the square of the distance in Gunter's chains by 800, the quotient will be the depression in *inches* very nearly; the correction may be obtained in *feet* by merely taking two thirds of the square of the distance in miles. Another convenient form for making this correction for *any* distance is,—to add to the arithmetical complement of the logarithm of the diameter of the earth,* or 2.378861, double the logarithm of the distance in feet, the sum will be the logarithm of the correction in feet and decimals;—thus, for example—required the correction for curvature in 1350 feet.

Log. of 1350 feet	3.130334
	2
	6.260668
Add arithmetical complement of the log. of the diameter of the earth..	2.378861
Log. of .04361 feet	8.639529

But our preceding observations have been made without regard to what is termed "refraction," which is an optical deception, causing the position of an object to appear *higher* than it really is, except it be situated in the zenith. This "optical deception" is caused by the *density* of the atmosphere, which increasing as it approaches the surface of the earth, bends or refracts any

* The arithmetical complement of a logarithm, is what the logarithm wants of 10.00000, &c., and the easiest way to find it is, beginning at the left hand to subtract every figure from 9 and the last from 10. Thus the diameter of the earth is 41,796480 feet, the log. of which, 7.621139, subtracted from 10.00000, &c. gives 2.378861, the arithmetical complement.

particle of light falling obliquely on it, more and more towards the perpendicular as it approaches the surface; consequently, the rays proceeding from any object describe a curved track, concave to the earth's exterior.* Every object invariably appears to lie in the direction of a tangent to such curved track at the point of observation, except, as we have before observed, it is situated in the zenith, when it is wholly free from refraction;—but, when situated near the horizon, refraction is at its maximum. Perhaps this matter will be rendered more plain by the aid of the following diagram. Suppose A B to be a portion of the earth's surface, G H the upper boundary of the atmosphere, S a star, P the place of the observer, and Z, his zenith; C D, E F, and G H, are assumed boundaries of the strata of the atmosphere, each of different density. A ray of light then proceeding from S, and impinging on the atmosphere at K,—where it encounters a denser medium than in its pre-

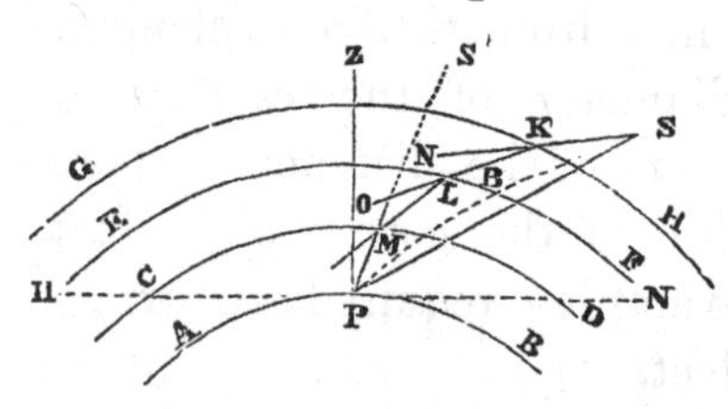

* The rays are occasionally affected in an extraordinary manner, sometimes laterally and at other times convex to the earth's surface instead of concave,—whereby objects have appeared depressed instead of elevated. Sometimes it has also been observed that the rays were affected in both ways at the same time, (*i. e.* elevated and depressed) appearing double, with one image erect and the other inverted; but this only occurs when the observed object is near the horizon and under peculiar circumstances of the atmosphere. In the account of the Trigonometrical Survey of England, a case of this kind is mentioned. "In measuring the base on Hounslow Heath—observes the narrator—we had driven into the ground at the distance of 100 feet from each other about 30 pickets, so that their heads appeared through the *boning* telescope to be in a right line; this was done in the afternoon. The following morning proved uncommonly dewy, and the sun shone bright, when having occasion to replace the telescope, we remarked that the heads of the pickets exhibited a curve, *concave upwards*, the farthermost picket rising the highest, and we concluded that they were not properly driven till the afternoon, when we found that the curved appearance was lost, and the ebullition in the air had subsided."

vious path, will be deflected in the direction K L; at L entering a still denser medium, it will be further deflected in the direction L M; similar effects taking place throughout its path, until at length it enters the eye of the observer at P, in the direction M P. The ray whose path we have just traced, is therefore not a straight line, but broken into numerous parts; and if we suppose the several degrees of density which we have assumed the atmosphere to be composed of, to be indefinitely increased, the path of the ray may be considered as curvilineal; its course would then be represented by the dotted line S B P, concave towards the earth's surface, and it would enter the eye of an observer in the direction of a tangent to that curve. The star S, would therefore appear to be at S′, and the angle S′ P S, would represent the refraction. Precisely similar results take place with regard to the rays of light by which terrestrial objects are rendered visible, only that the refraction is greater, consequent on the rays passing more obliquely through the denser portion of atmosphere.

The form and magnitude of the curved track described by a ray of light in its passage through the atmosphere, has been a subject of investigation with numerous scientific men; but as it varies with every state of the atmosphere, results obtained at one time can be rarely if ever applied at another, with any regard to accuracy; it may, however, be considered of regular inflexure, in form approaching an arc of a circle, and in all cases may be estimated either in terms of the curvature, or of the horizontal angle,—termed the arc of distance. When the atmosphere is in a *mean* state it may be estimated in the former case at $\frac{1}{7}$th of the curvature, and in the latter, at $\frac{1}{12}$th of the arc of distance, or angle subtended at the earth's centre. But in all extensive geodesical operations,* where the effect of refraction requires to be estimated; contemporaneous angles should be observed from either station, and the necessary corrections computed.† The variation in refraction which is observed in the atmosphere at different times, is generally produced by changes of temperature; cold condensing the air and increasing the refraction, while heat expands or rarefies the air—diminishing refraction. From this cause refraction is greater in cold than in warm weather; consequently it is less in the evening than in the morning, except under peculiar circumstances. Humidity of the air is *said* not to produce any sensible effect on its refractive power.

The correction for curvature and refraction we have computed and appended in the form of a table at the end of our treatise;—that for refraction being taken for the *mean* state of the atmosphere or $\frac{1}{7}$th of the curvature.

The examples which we gave of the effect of curvature at pages 4 and 5 in depressing the position of an object, will therefore be modified by the effect of refraction. The first example will therefore stand thus:—

Computed amount of curvature................	16.6750	Feet.
Deduct for refraction $\frac{1}{7}$th of curvature.........	2.3821	
	14.2929	
Add height of instrument above surface as before	4.5000	
True difference of level	18.7929	Feet.

* The effect of *ordinary* refraction is to alter the place of an object in a *vertical* plane, but it does not affect the azimuth;—therefore in geodesical operations where the sides of the triangles are not sufficiently large as to require to be treated as spherical triangles, or the heights of the trigonometrical points are not required, both refraction and curvature may be altogether neglected.

† For the method of computing this correction, consult Vol. I. "Trigonometrical Survey of England," Lieut. Frome's "Outlines of Trigonometrical Survey," or Woodhouse's "Trigonometry."

And the second example will stand thus :

Computed curvature for 1350 feet	.04361 Feet.
Deduct for refraction $\frac{1}{7}$th of curvature	.00623
True difference of level......................	.03738

Before concluding this chapter we think it proper to inform the reader that in the *ordinary* operations of levelling, the effect of curvature and refraction are wholly omitted in the calculations, except under very peculiar circumstances,—as the correction if attempted, would be so extremely small, as to be quite inappreciable, without the graduations on the levelling staff were carried to at least two or three places of decimals further than at present, and which would of course require the power of the telescope to be greatly increased. This would introduce a much greater number of figures in the calculations, and so possibly give rise to errors in that way; but as this part of our subject will be fully discussed in a more advanced part of our work, we shall at once dismiss it, and introduce the reader to the practical details of levelling as contained in the following chapter.

CHAPTER II.

LEVELLING.—THE GREAT IMPORTANCE OF THE SUBJECT.—VARIOUS KINDS OF LEVELLING INSTRUMENTS EMPLOYED.—THE SPIRIT-LEVEL AND STAFF.—DESCRIPTION OF LEVELLING OPERATIONS.—METHOD OF COMPENSATING FOR CURVATURE AND REFRACTION.—EXAMPLE IN LEVELLING WITH REDUCTION OF LEVELS.—THE DATUM LINE. —LEVELLING BY BAROMETER.—LEVELLING MACHINES.

THE art, or rather science, of levelling is of the greatest practical utility, more so, perhaps, than any of the multifarious callings which the science of construction calls into play. Without it the artificer's skill would be almost useless, and the talent of the architect or engineer would frequently be wasted in designing impracticable structures. Important public works have often been *designed* upon erroneous data in this respect, which it would have been quite impossible to carry into execution,—for instance:— The section of one of the projected lines of railway to Brighton, and commonly known as "Cundy's Line," was so incorrect, that had the "gradients" been secured which were laid on the section, it would have involved miles of cutting of from 150 to 200 feet in depth. Another instance within our knowledge is that of a projected canal which was confidently stated to be practicable— but to procure a sufficient supply of water, such a work was required as the mind of man never conceived—an aqueduct of several miles in length and of greater vertical dimensions than the preceding cutting! Again, a projected line of railway through Kent was two or three years since surveyed in detail, and brought into parliament, where the advocates for an act of incorporation, were compelled to admit of an error existing in some portion of their section, of 50 feet!—but yet they persisted in calling it, not only a practicable line, but superior to any other before the House. Now this error of 50 feet might probably have made all the difference between a practicable and impracticable line.* But there have been so many instances of this kind,—to say nothing of railway excavations and embankments not only projected, but executed on so low (wrong?) a level as to have ultimately required filling in, or raising considerably, that we might fill a moderate volume with their enumeration. Enough is it for us to observe, that these failures (causing an immense waste of money) have resulted from erroneous levels. The vast importance of the subject we think will therefore be readily conceded, and should we be considered at any time too minute or prolix in our directions, we trust our readers will excuse us, bearing in mind that we do not profess to inform the *initiated*, but to instruct the ignorant on one of the most delicate and cautious operations which the engineer or surveyor is ever called upon to perform.

The operation of levelling may be performed in various ways, calling into use instruments widely differing in construction and application, but all depending on one principle—gravity. We may mention incidentally, the plumb-line, on which the action of the mason's level depends; but such an instrument would require, in a theoretical point of view, a plumb-line of almost indefinite length to give *accurate* results. Its construction also involves so many practical difficulties, as altogether to require its rejection, except for the most ordinary purposes.

The next instrument in point of importance is the water or fluid level—depending for its action on the

* The *probable* cause of this discrepancy will be hereafter explained.

same principle as the plumb-line. If a bent tube is partly filled with water, and the atmosphere allowed freely to communicate with the fluid at either end—the surface will stand at the same level in both arms of the tube; therefore a line produced from the surface of the fluid in both arms, would be a line of level, and if *produced* sufficiently far (which is impracticable) would require a correction for sphericity. But if the two arms containing the fluid were ever so far distant, (providing the density of the atmosphere was the same at both places) no correction for sphericity or any other cause would be required, but both surfaces would indicate the true level, and be equidistant from the earth's centre. A very pretty application of this principle has long been in practice among French engineers, which may be described in a few words;—at each end of a metal tube of moderate length are fixed short glass indicator tubes, in a vertical position, opening upwards into the atmosphere, and downwards into the metal tube. If this metal tube is set up in nearly a horizontal position, and water poured in at one end, it will immediately rise to the same level at the other; we might then so place ourselves that our eye would be on the same line or level, as both surfaces of the fluid, and at the same time we might easily fix an object also in the same line or level, *beyond* the instrument, so that we should have four points on the same level line—*i. e.*, the surfaces of the fluid, the point of sight, and the fixed object beyond. Now, if the height of the observer's eye in such a case was just 4 feet above the ground, and the surface of the fluid in the glass tubes was elevated 6 feet, it is evident that where the observer stood, the ground would be 2 feet higher than that where the instrument was planted. Again, if the object fixed beyond the instrument from the point of observation, should be 8 feet above the surface—the ground there would be 2 feet lower than at the instrument, and 4 feet lower than at the point of observation. We have introduced this short example to show on what simple calculations levelling operations are based;—but to proceed.—The instrument which we have just described is rendered much more facile in practice by colouring the water with some chemical admixture, which at once renders its surface distinctly visible.

Sometimes a float is applied to each end of this instrument, which, rising with the fluid, determines the line of level with still greater facility, and any slight contrivance might be applied to its upper part to render the operation much more delicate and correct;—such as a horse-hair strained horizontally on each float, and placed transversely to the tube's length, by which means the line of level might be defined with very minute accuracy.

A great improvement on the common water levels which we have been describing, has recently been made by a Mr. Brown, for which he has obtained a patent. It consists in the affixing of a flexible in place of a metal tube to the indicator glasses, which improvement renders this instrument not only more portable, but applicable in cases where the previously described instruments would be useless,—as where the points of sight, or rather of observation, are hidden, the one from the other, by some intervening obstacles and in many other situations. It is not at all requisite that the hose or tubing of this instrument should be *below* the level of the water in the glass indicators; but on the contrary it—and of course the internal column of water—may be elevated to such a height as the pressure of the atmosphere will balance, which, under ordinary circumstances, is about 34 feet. Thus the difference of ground line on opposite sides of a wall may be easily ascertained with this instrument, by simply passing the hose over the wall, and measuring the depression below the water surface, as shown in the indicator; the water surface on the other side being level with the ground. In a similar manner it will be found efficient in excavating foundations, where from the necessity of commencing detached portions of a structure at a time, it is often difficult and troublesome for the workmen to keep the bottom level the same throughout; and in the throwing out of bridge-pits, the setting of heavy masonry and centerings, the erection of machinery, and such like purposes, it will be found particularly serviceable. The accompanying wood-cut represents this instrument in

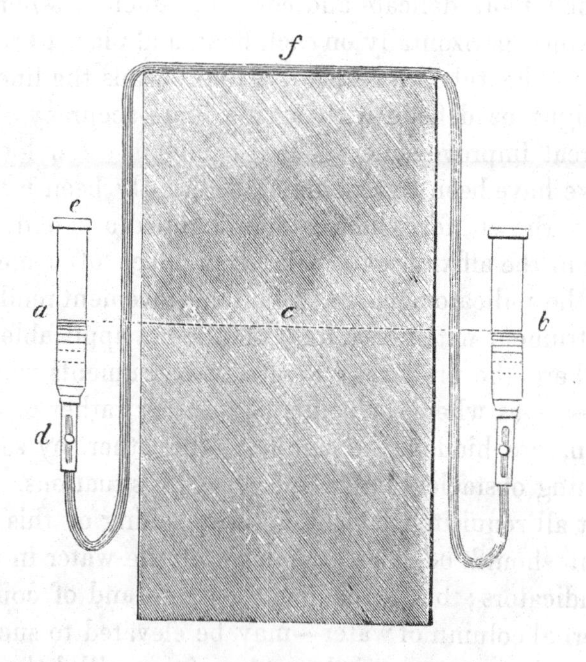

use.—*c* is a wall and *a f b* the "hydraulic level" passing over it by means of its flexible tube *f;* the surface of the water in the indicator glasses *a* and *b* denotes the horizontal or level line on each side of the wall. Great care is necessary in expelling the air from the hose when filling it with water, otherwise the action of the instrument will be very imperfect; the thorough expulsion of the atmosphere is in fact the only adjustment required, but on which its accuracy altogether depends. This adjustment is performed by putting the column of water in motion, by elevating one end, until all the air bubbles in the hose are expelled, and the water stands in both indicators exactly at the same height when placed side by side. To confine the water in the hose when not in use, a stop-cock *d* is applied, the screw-cap *e* is to prevent the water running out when one end is depressed,—as when taking the level of ground line on opposite sides of a wall, but it is so applied that the action of the atmosphere remains unimpaired. The hose in which Mr. Brown's improvement consists, is made in 5 feet lengths which screw together; so that any required length may be obtained. We have been informed that this instrument with a 100 feet hose has been used on the Great Western Railway for laying down and adjusting the rails, and the inventor insists on its applicability to the obtaining of the profile of a country, in place of the ordinary operation of levelling by spirit-level and staff. But we very much doubt not only the advantage of using the "hydraulic level" for such operations, but in being able to arrive therewith at even an approximate correct result. The length of time which such a column of water will continue to oscillate, when once put in motion, is very great, and the uncertainty of the result, if noted while in such state, easily imagined,—to say nothing of the difficulty experienced in transferring the levels, and the unscientific mode of obtaining the value of their difference; only to be ascertained by bringing the graduated staff (which must be employed) in contact with the water line in the glass indicators, at each remove of the hose.

Another instrument which has been employed for levelling, but we believe only by military engineers, is the "Reflecting Level," the action of which must be obvious from its name. A small square piece of plane mirror, fixed in a metal frame, comprises this instrument, and although it cannot be employed for extensive operations, will yet give very correct results on a small scale. Its principle and mode of application we will explain in a few words. The rays proceeding from an object in front of a plane mirror are reflected as if diverging from a point the same distance behind as the object is in front. Now if the frame in which the mirror is fixed, be suspended so that its face is vertical, the eye and its reflection will be on the same level line, and any object in front of it, which coincides with this line of sight, will be reflected on the same level. By suspending the mirror diagonally, with a fine wire or thread strained horizontally across its face, a zero will be obtained from which the elevation or depression of the ground in any direction may be ascertained,—the observer simply measuring the height of his eye above the ground when reflected in the mirror, the difference between which and the height at which the mirror is set up, being the difference of level; or it may be otherwise obtained with the assistance of a graduated staff with a sliding vane, which is recommended always to be used with this instrument. A small portion of one angle of the mirror is cut off, across which the horizontal wire passes, by which means the level line can be prolonged *behind* the instrument, and the length of sight thereby increased; the instrument also by this arrangement admits of very delicate adjustment, but which merely consists of such an alteration of its centre of gravity, and consequent change of position in the face of the mirror, that when *reversed* it reflects an object *on a rear line*, produced from its previous position.

For levelling by angular measurement there have been many instruments employed, such as the "graduated triangle or quadrant," the angle with such an instrument being determined by a plumb-line depending from the centre of the arc; but of course by its use much more complex calculations are introduced than are necessary when merely a series of horizontal lines are traced, as with all the former instruments which we have described. The same remark applies to the use of reflecting instruments, as the quadrant, sextant, &c., as also to the theodolite.* A very old method of obtaining differences of level by "proportion" is with what are termed "boning rods," which are merely a few wooden rods of an equal length having a rectangular cross piece fixed on one end, and which will be well understood by the letter T. The mode of operation is this: having determined the difference of level by any means for a distance of say 50 feet —which suppose to be 2 feet; then by placing a boning rod at each station, and looking along the T heads of the two rods, a third may be placed at any distance, having the same inclination to the horizon as those previously set up. Now supposing the distance between the second and third rods to be 500 feet, we shall have by proportion

$$\text{As } 50 : 2 :: 500 : \frac{500 \times 2}{50} = 20 \text{ feet};$$

and this system may be continued for any distance that the inclination of the ground continues the same. By

* The quadrant, so called, from its being the fourth part of a circle, the sextant the sixth part of a circle, and the theodolite, from two Greek words, signifying "to see," and "distance." The process of levelling with the theodolite will be found in a subsequent chapter.

a little attention also to the arrangement of the rods, this kind of levelling may be continued over ground of *varying* inclination, without having recourse to any other instrument, or means to determine the inclination of the ground except at the commencement;—for instance, supposing, in continuation of the example which we have given, the inclination of the ground should change at the point where the third rod was set up; then by measuring out a further distance and setting up a fourth rod, observing the height above or depth below the ⊤ head intersected by the line of sight, the rise or fall from the third to the fourth would be ascertained, which could then be employed proportionally in ascertaining the level of the succeeding inclination, and so on for ever so great a distance. A modification of this kind of levelling has been proposed in conjunction with an instrument termed a "Cambrian level," and which we shall proceed to describe. The difference from the preceding method consists merely in the first term of the proportion being determined with the instrument itself, or rather it is set to the inclination of the ground, which, being traced and measured for any distance, the difference of level is deduced as before. The construction of this instrument is exceedingly simple, being nothing more than two arms of wood or metal similar to a carpenter's rule; the lower arm is graduated into inches or any other divisions, and along the upper side of which moves a stud of metal in the shape of a right-angled triangle, its vertical height being exactly equal to one of the divisions on the arm. Of course when this stud is set at 20 divisions from the junction of the arms,—the lower one being level, and the upper one resting on the stud, an inclination of 1 in 20 will be given, and so on for any other inclination. It will not fail to be perceived that the correctness of such an instrument will depend on the magnitude of the graduations and accuracy with which the lower arm is set level, which requires an arrangement of the plumb-line, water-level, or other contrivance as by to the preceding method, and therefore, we cannot perceive any advantages that would result from its use, beyond what is within our reach by the use of the boning rods; indeed we should much prefer the latter, from each term of the proportion being in whole numbers, instead of in fractions of the required resultant, as well as from their superior cheapness and simplicity. For further remarks on the use and application of "boning-rods" we must refer our readers to that part of our treatise which relates to the setting out of works.

Having now detailed all the *ordinary* instruments which have been called into use in the operation of levelling, with the exception of the spirit-level,* we will devote the remaining portion of this chapter to a short description of that instrument, with the method of conducting simple levelling operations: reserving for an advanced chapter such remarks as may be necessary as to its construction, adjustments, &c. The ordinary spirit-level consists merely of a telescope with cross wires fixed within its field of view, a glass tube *partly* filled with spirits of wine (hence its name) placed parallel with the optical axis of the telescope, and a stand on which to fix the instrument when observing; having simple mechanical contrivances for setting the whole level, at the same time permitting the telescope to be turned about in any direction. Used in conjunction with the spirit-level, is a graduated staff, by which the changes of level are measured; for it is evident that with the spirit-level alone, although differ-

* A more proper name for this instrument would be the "air level," for it is the air bubble and absence of the spirit which points out the horizontal line.

ences of level might be observed, the value of such differences could not be ascertained without having recourse to a "graduated staff;" for if a carpenter's rule was afterwards applied to ascertain the value of such differences, it would represent such an instrument, clumsily applied. The levelling-staff, until a recent period was made with a sliding vane, moveable up and down a rod graduated into feet and inches, or decimally into feet; the position of the vane in regard to the ground line, being easily determined by the graduations on the rod. The vane was considered a necessary appendage to the staff, as affording a conspicuous mark when viewed through the telescope at considerable distances, and from the small power of the glasses applied to levels a few years since, was essentially necessary. But within that period, levelling telescopes have been constructed of such superior power and brilliancy, that by merely rendering the graduations a little more conspicuous, the observer is enabled distinctly to read them, at as great a distance as it would be prudent under any circumstances to note the position of the sliding vane. The construction and use of the two essentials to levelling being understood, to wit, "the spirit-level and staff," we will proceed to explain the method of using them in taking "the profile" or section of a piece of ground. The act of levelling is commonly termed "taking a section," and thereby is exceedingly apt not only to give learners a wrong notion of what is to be done, but even in the minds of persons having some slight knowledge of the matter, it creates a confused mystification, frequently, we are satisfied, leading to very disastrous consequences. If we could conceive that we were *surveying* the ground, and the horizontal line traced by the spirit-level, to be the base line, the rises or falls of the ground, offsets,—we should then have not only a correct, but a very clear notion of the operation of levelling; in fact the only difference between surveying and levelling is, that the former is measured on the horizontal plane, and the latter on the vertical. A section is not determined by the operation of levelling, which merely implies the ascertaining of the variations in the ground-line or surface, and therefore might strictly be termed "profiling." By the after operation of boring, or by observation of the out-crop, the strata for a certain depth may be determined—then, but not till then, a *section*, other than imaginary, can be drawn.

We have mentioned that the spirit-level, when in use is set up on a stand, so that on looking through the telescope the line of sight is horizontal; but this is only correct when the observer looks at the transverse wire which intersects the field of view; if he looks the least above or below it, the line of sight will be oblique to the horizon. It is clear, then, that when examining a level staff through the telescope, that that graduation which is *bisected** by the transverse wire will be on the horizontal plane, and none other; if then the axis of the telescope is elevated 4 feet above the ground, line, and the wire bisects the graduation at 6 feet on the staff, the ground where the staff is situated will be 2 feet lower than where the instrument is planted. We may now safely proceed to an example:—Let A B be the extent of a piece of ground, the profile of which is required, the several variations in the surface being shown at *a b c*. Plant the spirit-level at A,

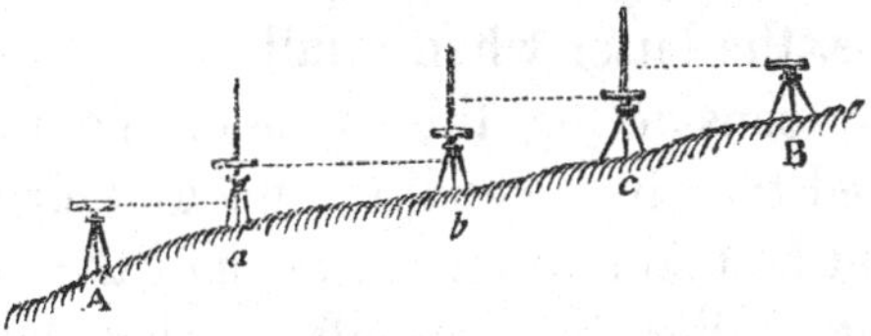

* See observations on levelling staves.

carefully set it horizontal with the object glass directed towards *a*, and measure the height of the centre of the telescope above the ground, which suppose to be 4 feet; then send forward the staffman with directions to hold up the staff where the inclination of the surface changes, —as at *a*, and read the graduation on the staff, which say is 3 feet, this sum deducted from the height of the instrument, will give a *rise* of one foot from A to *a*. The spirit-level is then to be removed to the spot where the staff stood, and again set up horizontal, with the object-glass directed towards *b;* now supposing it to be set up at the same height as before, and the staff at *b* to be bisected at 2 feet, then there would be a *rise* of 2 feet from *a* to *b;* making *b* 3 feet higher than A.

The process which we have just described was the system of levelling generally pursued up to a very recent period, the level and staff being continually set up at short distances from each other—and the difference between the height of the telescope and the bisection of the staff entered as the difference of level, *plus* or *minus*, the former sign being appended to the observation when a rise, the latter when a fall.

By considering this mode of operation, it will be perceived that in ascending, no greater difference of level could be taken in one sight than the height of the instrument; therefore, a staff of that height would have been sufficient under such circumstances. But we have in remote parts of the country often seen surveyors thus levelling with staves of 10 or 15 feet in length, and wondered that that circumstance had not suggested a better mode of operation, so that at least the level and staff might have changed places, the former always occupying the ground of *superior* elevation; by which means a difference of level equal to the whole height of the staff, *minus* the height of the instrument, could be taken in one sight. We may illustrate this matter by reference to the last diagram;—suppose we were about to commence operations at *b*, the ground rising towards B, we should set up the staff at *b*, and the spirit-level on the *superior* ground at *c*, looking backwards to the staff at *b*, which suppose was bisected at 10 feet—the height of the instrument being the same as before, viz., 4 feet—we should have a difference of level of 6 feet from *b* to *c*. The next process would be to remove the spirit-level to B, and the staff to the spot previously occupied by that instrument at *c;* and supposing the heights to be the same as before, that is the bisection 10 feet, and the height of the instrument 4 feet, we should have a rise of 8 feet from *c* to B, making the total rise from *b* to B 12 feet.

It must be evident to the most superficial reader, that the preceding methods of taking the profile of ground are open to very serious objections; for in the first place, the trouble and uncertainty of determining the height of the telescope above the ground is very great; added to which is the necessity of setting up the level or staff on removal at the exact point from which the height of the telescope was measured, or staff placed, otherwise an error equal to whatever difference of level may exist between the points taken will be introduced. Again, an error arising from *curvature* and *refraction* is insured, the amount of which will depend on the length and number of the sights; and in the next place if from any slight mechanical defect—and all instruments have such defects in a greater or less degree—the axis of the telescope instead of being level, as supposed by the operator, should incline upwards or downwards, it is evident that each observation would be vitiated by the obliquity of the line of sight; which increasing proportionally with the distance to which a sight may be extended, as well as in the aggregation of the sights, must in a very short distance altogether produce so large an amount of error, as to render an operation of levelling, thus conducted, altogether useless, and sometimes exceedingly mischievous.

Some of our readers will then probably wonder in what manner correct results may be obtained in levelling operations, for in our remarks in the last paragraph, we hinted that although corrections for curvature and refraction should be applied,—still, from the mechanical defects of all levelling instruments, accuracy could not be obtained. We here again beg to repeat the observation, and now proceed to explain the simple method by which, not only the error arising from curvature and refraction may be neutralised, but that also arising from mechanical defects in the instrument. We have before observed that the spirit-level, in addition to its mechanical contrivances for obtaining horizontality, has the property of retaining that position when the telescope is pointed in any direction, provided the instrument is in proper adjustment. By this means then we have the power of destroying the *effects* of curvature and refraction, as well as that arising from the defects of the instrument. In the following example suppose *f g* to represent a portion of the earth's surface, with a spirit-level planted at *e*, and levelling staves at

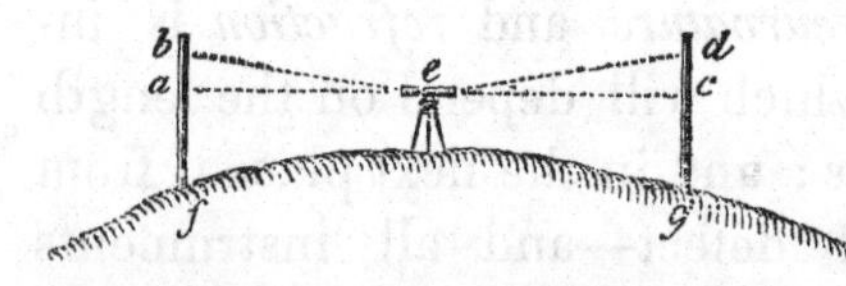

f and *g*, each one mile distant from *e*; then, if the spirit-level was set up truly horizontal, and wholly free from mechanical defects, the line of sight *e a* in observing the staff at *f*, would *not* represent the *true* level; but the observed difference between the height *e*, and the height *a*, would require such a correction for curvature and refraction as we have pointed out in our introductory chapter. But if without changing the site of the spirit-level, further than reversing or turning the telescope in the direction of the staff at *g*—which the construction of the instrument admits of without destroying its horizontality, we note the bisection of the staff by the line of sight *e c*, we should have an observation on each side of the spirit-level, both of which would be equally affected by curvature and refraction, and therefore the *true difference* of level between *f* and *g* would be obtained without applying any correction. By this system of levelling, which is termed "back and fore sight," it is not always necessary to set up the staves equidistant from the spirit level, even when the operation is of the most precise character; for if the staff *g* was set up twice as far from the spirit-level as the staff *f*, the requisite correction for curvature and refraction would be made on the excess of the square of *e c*, over *e a*.

Another and still greater benefit than the preceding, is derivable from the system of levelling by "back and fore sights," and which we may illustrate by reference to the last diagram. We will suppose that either from incorrect adjustment, or a mechanical defect in the spirit-level, the line of sight when taking the back observation should not be level, but inclined upwards, as *e b*;—now, by reversing the telescope on taking the forward observation, the line of sight *e d* would be elevated an equal quantity above the true level with the first; consequently, the correct *difference* of level would be given between *f* and *g*, notwithstanding the inaccuracy of the instrument. This fact the reader will do well to bear in mind, as on it depends the adjustment of all levels, and to which we shall have occasion to refer in a subsequent part of our work, descriptive of the several kinds of spirit-levels and their adjustments.

We now proceed to give a short example of levelling by back and fore sights. In the following diagram let the difference of level between A and I be required, and also the undulations of the intermediate surface, so as to be able to draw a profile of the ground to scale.

In the first place, set up the staff at the commencement A, then plant the spirit-level at as great a distance from the staff, within moderation, as the nature of the ground will admit, so that the line of sight does not intersect the ground line before reaching the staff, but still as near to the bottom of it as can be readily managed without a second levelling of the instrument. When this

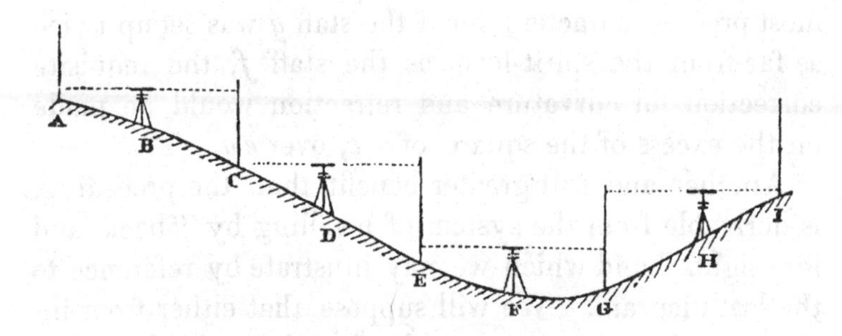

is accomplished, and the operator satisfied that his instrument will reverse without sensible alteration in the level, the reading on the staff at A is to be registered; the telescope is then to be turned in the direction of C, where the staff previously set up at A, is to be removed—or, to save time, a second staff is sometimes employed,*

* In extensive levelling operations where dispatch is desirable, it is quite common to use two staves, by which means some portion of the principal's time is saved; for, after reading the staff at the back station, instead of waiting until the man has advanced with it to the forward station, a man will be

and the reading entered as before. Suppose then the reading at A to be 4 feet, and at C 7 feet 6 inches, the difference, (3 feet 6 inches), will be a fall from A to C; the distance between which points is to be measured and entered opposite the difference of level, as is also the distance between each subsequent station where the staff is planted. The spirit-level is then removed from B to D,—a point as far beyond the forward staff as convenient, and accurately levelled as at first, when the bisection of the backward staff—which is not moved since the previous reading except to reverse the graduated face, is entered as 2 feet, and the forward staff E, 9 feet, the difference (7 feet), being a fall; to this is to be added the previous fall of 3 feet 6 inches, making the total fall from A to E, 10 feet 6 inches. In continuation of the operation the spirit-level is planted at F, the back reading to the staff at E being 4 feet, and the forward reading 7 feet, equal to a fall of 3 feet; which, added to the previous quantity, makes a total fall of 13 feet 6 inches from A to G. On the spirit-level being planted at H, the back reading was 6 feet, and the forward reading 2 feet 6 inches, equal to a *rise* of 3 feet 6 inches from G to I; which *deducted* from the sum of the depressions at G, makes the gross fall from A to I, 10 feet. This method of procedure might be continued for any required distance, adding to or subtracting from the previous reduced level,—or sum of the differences, according to the variation in the ground. If the difference of level between the extreme points A and I was only required, it would not be necessary to measure the distance between the staves, neither would it be requisite to reduce the levels of the several intermediate stations; as

there with the second staff, by which means the two sights can be taken almost simultaneously; and before the principal can set his instrument in advance of the forward staff, the previous back staff will be planted in readiness for a fore sight, and so on throughout the operation—see a subsequent Chapter.

the required result would be given by simply arranging the fore and back sights into separate columns, adding them up, and taking their difference, which would be the variation in level,—thus :—

	Ft. In.		Instrument at		Ft. In.
A =	. - 4 .. 0	- - - -	B	- - - - -	C = 7 .. 6
C =	. - 2 .. 0	- - - -	D	- - - - -	E = 9 .. 0
E =	. - 4 .. 0	- - - -	F	- - - - -	G = 7 .. 0
G =	. - 6 .. 0	- - - -	H	- - - - -	I = 2 .. 6
Sum of back sights	16 .. 0 feet.		Sum of fore sights		26 .. 0 feet.
			Deduct lesser sum . .		16 .. 0 ,,
			Difference of level between A and I - -		10 .. 0 feet.

But to be able to draw the section or profile of ground between the extreme points, it would be necessary to measure the distances between A and C, C and E, &c., as just explained; so that the staff being set up at each change of inclination, and the distances from staff to staff registered against the corresponding reduced level, the section could be plotted. But in such case the form of field-book would be similar to the following, the difference between the back and fore sights being in each case added to, or subtracted from, the preceding reduced level; but it would be necessary to add up the columns of the fore and back sights (as in the example above), to prove the correctness of the casting; and if their difference equals the last reduced level, it determines the accuracy of the computation.

Field Book.

Back Sights.	Fore Sights.	Red. Levels.	Distances in Links.	REMARKS.
		0 .. 0	000	Datum, level of ground at commencement.
4 .. 0	7 .. 6	3 .. 6	200	
2 .. 0	9 .. 0	10 .. 6	400	
4 .. 0	7 .. 0	13 .. 6	600	
6 .. 0	2 .. 6	10 .. 0	800	
16 .. 0	26 .. 0			
	16 .. 0			
	10 .. 0	Difference, the same as last reduced level.		

In this example we have the materials conveniently arranged before us for drawing the section; as in the column containing the reduced levels (taking the level of the ground at A as the datum, the meaning of which will be immediately explained) we have at 2 chains from A, a fall of 3 feet 6 inches; and at 4 chains (the distances being continuous) we have a fall of 10 feet 6 inches below our starting point. We thus continue the operation until the result of each observation has been plotted; first marking off the horizontal distances on the datum line, and then the reduced level or vertical answering to each distance. The method pursued in plotting a section will be immediately understood by reference to the following diagram, which represents the plotted section of the preceding example; the horizontal scale being two and a half chains to one inch, and the vertical scale twenty feet to one inch.

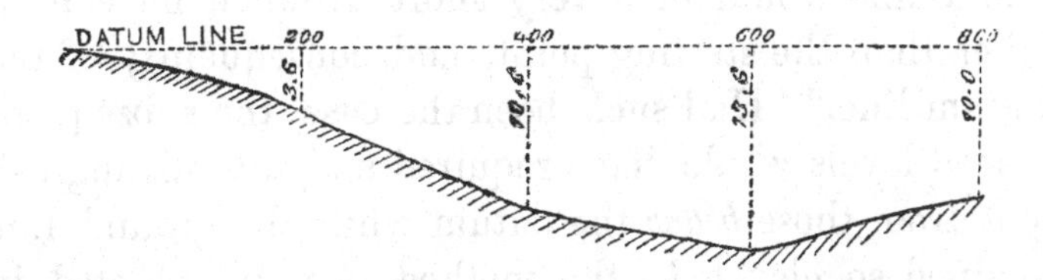

We may here observe that the vertical scale to which a section is plotted is always *different* to the horizontal; as if such an artifice was not resorted to, slight variations in the ground line would be imperceptible, and very extensive changes of level would be scarcely observed.

The word "Datum" in connection with a section, signifies a standard, a continued reference point or horizontal line concentric to the earth's exterior, to which all the calculations are referred;—thus, in the last example, it would be found extremely inconvenient in plotting the section, to have to draw the ground line at

E 7 feet lower than at C, at G 3 feet lower than at E, and at I 3 feet 6 inches *higher* than at G; instead of marking them all off as so many feet lower than the starting point A, which, in such case would be registered thus: "*Datum Line*" *level of starting point.* This plan of reducing all the heights or depths to a zero, or common standard, has also this additional advantage, that if by accident a wrong vertical dimension should be marked off, the error will not extend beyond that one point, whereas by the former method it would extend to each subsequent one, and render all such portion inaccurate; it will also readily be perceived that a dimension can be more accurately marked off at *once*, than if the same had to be done in two, three, or more operations. It will be easily understood on examination of the example we have given, that if the section had been continued beyond I, —the inclination of the ground remaining the same, the ground line would in a very short distance have been higher than the starting point, and consequently of the "datum line." Had such been the case, the subsequent reduced levels would have required a *sign* to distinguish them from those *below* the datum while the ground line remained so elevated; the method usually adopted in such cases is to place the sign + against all reduced levels *above* the datum, and the sign — against all those *below* it. But by a very simple expedient the necessity for such signs may be altogether avoided; for if the lowest level in any section was supposed to be 50 feet lower than the starting point, we might avoid having recourse to + or — levels by assuming the *starting point* to be 100, or any other number of feet *greater* than 50 above an imaginary line, which would be termed the datum; then all the reduced levels would be + or above the datum, and therefore no need of an algebraic sign. Equal facility is also afforded for the comparison of intermediate heights as by the previous method, for by rejecting 100 feet from any reduced level—their absolute rise or fall from the starting point would be given.

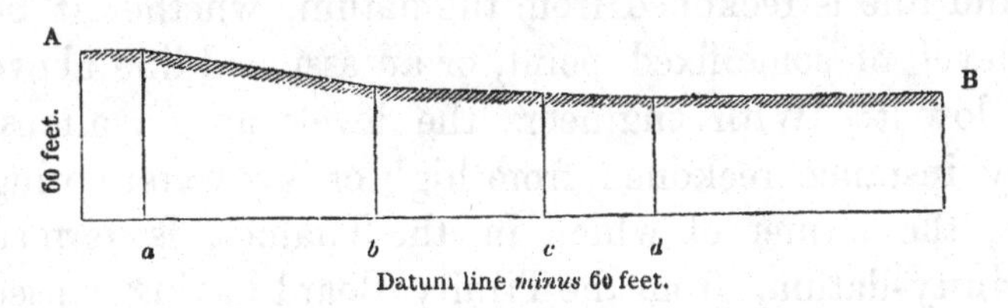

Datum line *minus* 60 feet.

The accompanying example will elucidate the matter:—The level at B, it will be perceived, is lower than at A, and suppose the intermediate levels between these two points to be required (commencing from A, and the amount of rise or fall not being known), it would be necessary in such case to fix on a datum line to commence or reckon from; in the example we have taken it at 60 feet below the level of starting point. The reduction of levels at the several intermediate stations is exhibited in the following column:—

Station A	60 feet above assumed datum			
- „ - *a*	58	„	2 feet fall	at *a*
- „ - *b*	45	„	15 - „	at *b*
- „ - *c*	42	„	18 - „	at *c*
- „ - *d*	41	„	19 - „	at *d*
- „ - B	42	„	18 - „	at B

In this example it was not necessary to choose any other datum line than starting point, all the subsequent levels being below it; but in the generality of cases the operator will not know the highest point until after his reduction of levels, and it will be a very rare case indeed, where the levels are commenced at the summit of a section. But even if such was always the case, it would be found an advantage to choose the datum *lower* than the *lowest* point on the section, for convenience of connecting it with other sections in continuity, which at a future time might possibly be required. We have already fully explained that every variation in the ground line is reckoned from the datum, whether it be the level of some fixed point, or an assumed line above or below it. With engineers the levels are in almost every instance reckoned from high or low water spring tides, the former of which in the Thames, is termed "Trinity-datum," from the Trinity Board having caused its mean level to be marked at Sheerness and other places, from which zero the flood and ebb of the tide is registered.* In some parts of the country local standards are referred to, as at Liverpool, all the railways and canals in that district being referred to the average of low water mark. Such an arrangement is of the greatest advantage in enabling every engineer, employed in that district, to reduce his levels with certainty to the established datum. Further remarks on the datum line will be found in a subsequent part of our work.

In concluding this chapter we may just refer to some methods which have been proposed for levelling—and in some instances practised, widely differing from any of the preceding. The method of determining heights with the barometer, by the temperature of boiling water, and various other contrivances for the purpose of measuring the density of the atmosphere, we can only allude to, as such methods of levelling can never be employed

* It would be of the utmost benefit if a similar arrangement was made at all large seaport towns, and the actual difference, either in the low water or mean level of the sea established, not only in an engineering point of view, but as a means of throwing much light on a subject now but little understood;—we mean the action of the tides, which, although at present carefully recorded at most places, yet, so far as we know, the relative levels not being known, the most interesting part of the investigation is neglected.

with any degree of certainty.* Their only recommendation consists in the extreme rapidity with which comparative heights over a country may be taken, and thereby a rough section obtained; but where a distant approximation to the truth is only required, we should be more disposed to place confidence in a series of angles of elevation and depression taken with a good theodolite, the base in each case being obtained from the ordnance or other correct map of the country. Various kinds of levelling machines have also been proposed—which being drawn along a road, describe its profile. The actuating principle of all such machines, it is evident, must be the pendulum, which with machinery for unwinding and winding up the paper, and regulating the draught of the datum and surface lines, comprise this instrument. Several objections may be made to the use of such machines, among which we may mention the unsteadiness of the pendulum, consequent on the unavoidable shocks to which it must be subject while being drawn along an ordinary road; and the datum line in no case representing the horizontal distance—except where the ground is level—but the inclined surface over which it is drawn: consequently the inclination is less than it really is; but as it cannot be made available for taking sections across an enclosed country, and in fact for no other sections than of good roads—the levels of which throughout the kingdom are no doubt already recorded—we see no chance of its being brought into use; it will therefore be useless proceeding further with these remarks.

* One-tenth of an inch in the height of the mercurial column is equal to about 90 feet of altitude, which difference has been observed at the same place within a very short period of time, and at places on the same level within a short distance of each other at the same instant; it is evident its application must be attended with such uncertainty as to be useless in engineering operations, though under favourable circumstances near approximations to the truth have been obtained.

CHAPTER III.

DIRECTIONS FOR TAKING TRIAL SECTIONS.—EXPLANATION OF BENCH MARKS.—EXAMPLE OF TRIAL LEVELLING.—COMPARATIVE SECTIONS.—REDUCTION OF LEVELS.—INTERMEDIATE SIGHTS AND DIFFERENT FORMS OF REGISTRY.—CHECK-LEVELLING.—EXAMPLE OF LEVELLING FOR A RAILWAY.—ERRORS COMMON TO LEVELLING OPERATIONS.

HAVING gone through the various methods which have occasionally been resorted to for obtaining differences of level, we shall for the present dismiss all such with the exception of that by the spirit-level and staff; the use of which, in taking sections under a variety of circumstances, we proceed at once to explain.

The first operation in the way of levelling—say for a projected road, is to take a *trial section* in the direction supposed most favourable for the object in view. To proceed properly with such an operation, a good map of the country should be procured, and the proposed line of road laid down thereon; which in most cases passing through certain well-defined points, such as a cross road, a farm-house, the angle of a wood, or other prominent features, will enable any person of common observation to trace it out with facility. The persons required in taking a trial section, are the engineer or surveyor to observe and register the levels; a careful assistant—who may be an ordinary person of common abilities, to chain and give the distances to the principal, and two other assistants, one to draw the chain and the other to hold the staff. On the principal being satisfied that his level is in adjustment, he proceeds a short distance along the proposed line to explore, so as to enable him to direct his assistants in the exact direction in which they are to chain; but if the direction of the line should be well defined, such an exploration will not be necessary, and the operation of chaining and levelling may be at once commenced. Where the direction is not well defined, it is usual for the principal roughly to range out the line with poles and flags; a much better and cheaper method which we often practise, is to range it out with laths, which are easily carried, and form very conspicuous marks. If the commencement of the section should be at a road, as in the generality of cases, the chainage should be commenced and carried forward from the centre; which point as far as the section is concerned, is the first station to be observed, although it is not generally the case, from the necessity of establishing *Bench marks* at the termini of all levelling operations. The use of the bench mark and explanation of the term we proceed to give before advancing further with our directions for taking a section. Suppose the levels to be commenced at the centre of a road, or on one side of a field, and that after levelling the required distance the difference of the extremes should be 100 feet; and that at some subsequent period the levels were required to be taken in another direction commencing from the first station, for the purpose of ascertaining the relative height of another terminal point. In such case it would be requisite to commence at the exact level where the difference was 100 feet; for supposing operations were commenced at a point where the difference was 95 feet, but which the operator supposed to be the same level as the commencement of his preceding operations, it is evident all the subsequent levels would be 5 feet in error; and it should be borne in mind that there can be no verification—as in surveying, except by going over it again. But if at the commencement of the first levels, a remarkable stone was just flush with the surface of the ground, and on which the staff was planted, it is evident that no such error would occur on resuming and carrying forward the operation at a subsequent period, from the identity of the stone, and its surface not having changed its level since the preceding operations;—in such case the surface of the stone would be a *bench mark*. Now it can but rarely happen that such a bench mark exists at the commencement of a section; it therefore becomes necessary for the operator to look about for a convenient object on which to *bench*. If such cannot be found close to the spot and on the same level, it may probably at a few feet higher or lower, which, if the difference is noted, will afford the same ready facility of reference and comparison as in the former case. The objects usually chosen for bench marks are hooks of gates, boundary posts, notches or benches cut in trees, (hence their name) &c., and require to be marked in a peculiar manner for the purpose of identity at a future period; the necessary remarks for such purpose should be made opposite the entry of the level, and it will always be found advisable to sketch the shape and position of the bench mark.

This point being then determined, the principal sets up his spirit-level in such a position as to command a view of the staff when placed on the bench mark, at the commencement of the section, and also at an advanced point on the line; the distance of which from the commencement is determined by the assistant, and registered by the principal opposite to the corresponding level. The spirit-level is then removed, and set up *beyond* the staff, and the back sight registered; the chainage in the mean time being carried forward for as great a distance as the sight extends, which of course in the generality of cases, will depend on the undulation of the ground. The staff is then to be removed from its last position and set up at the forward station, the reading of which, with the chainage, being registered by the principal as before. In a similar manner is the operation continued for any required distance, bearing in mind that the staff is to be planted at as near equal distances from the spirit-level as the ground will permit, although it is not absolutely necessary that such should be the case to arrive at accurate results, except under peculiar circumstances, of which we shall have occasion to speak presently.

The object in taking trial sections being to ascertain generally the best of several lines, it is not necessary to take the difference of level or to show on the section every *trifling* variation in the ground, but only the principal features of each particular line; as in all cases of trial sections, the levels are again taken over the particular line decided on, when greater minuteness is of course observed than in the preliminary operation.

The trial sections in part represented at plate 8, were taken for the same line of railway, although some by the different routes taken are much longer than others. The method of plotting those sections for the purpose of comparison is obvious; they are all based on the same datum line, the same bench marks being referred to at the termination, as well as the commencement of each line; by these means the accuracy of the levellings were tested, which, if practicable, *should always be done.* It was not known at the time of taking the sections how much above Trinity datum the point of commencement was, no levels or bench marks having been determined; consequently a bench mark was left, and assumed as 100 feet above it, to which all the levels were reduced and plotted.

It was afterwards ascertained that the height of the

B. M. was 115 feet above the Trinity high-water standard, or 15 feet higher than that assumed; it only therefore required a line to be drawn on the section 15 feet below that adopted, to give the correct datum; and by adding the difference of 15 feet to any reduced level in the field book, of course the exact height of that point above the Trinity standard would be given without reference to the section. Also by taking two or more levels reduced to the same datum, and deducting the one from the other, the rise or fall between such points will be ascertained with as much facility as if each level was registered as so much above or below the starting point, or of that immediately preceding.

By the method which we have adopted, of tinting the surface of each section a different colour, their merits are at once determined without confusion; but the field book of the section tinted brown is only given as being sufficient for our purpose. The page at which the levels are commenced should always be headed somewhat in the following manner:—Trial Levels for the —— railway, canal, or road, as it may be, from (the name of the place commenced at, to the place of termination), with the date, name of the employers, and such other particulars as may be considered necessary.

Field Book.—Datum assumed* 100 *ft. above Trinity High-water : Distances in Chains.

Elevation.	Back Sight.	Fore Sight.	Depression.	Total Rise.	Distance.	REMARKS.
				100.00	0,00	B. M. above assumed datum on root of tree, level with ground at commencement.
7.86	9.20	1.34	......	107.86	2,00	
10.89	11.32	.43	....	118.75	3,70	
9.64	11.24	1.60	......	128.39	6,00	
......	7.80	10.68	2.88	125.51	10,00	
......	2.76	7.16	4.40	121.11	18,00	
7.27	9.22	1.95	......	128.38	28,00	
35.66	51.54	23.16	7.28			
7.28	23.16			100.00	......	First reduced level to be deducted from last.
28.38	28.38			28.38		

Elevation.	Back Sight.	Fore Sight.	Depression.	Total Rise.	Distance.	REMARKS.
				128.38	...	
....	4.38	7.06	2.68	125.70	37,50	Side of Road. [width 100.
....	4.10	5.30	1.20	124.50	38,00	Centre of T. P. Rd. to London,
....	5.30	10.78	5.48	119.02	38,50	Side of Road.
2.30	7.35	5.05	....	121.32	46,00	
3.32	11.36	8.04	....	124.64	55,00	
....	4.55	10.00	5.45	119.19	61,00	
....	2.20	10.43	8.23	110.96	63,50	
....	.42	11.90	11.48	99.48	64,80	
....	.49	11.46	10.97	88.51	65,50	
....	.90	15.00	14.10	74.41	67,00	
14.42	15.00	.58	....	88.83	70,00	
9.41	10.84	1.43	....	98.24	73,00	
9.18	11.34	2.16	....	107.42	75,00	
7.85	10.75	2.90	....	115.27	79,00	
6.88	10.88	4.00	....	122.15	83,00	—1 Mile.
	4.00	9.36	5.36	116.79	7,00	
....	1.32	8.86	7.54	109.25	B.M.	Top of style.
....	.70	3.00	2.30	106.95	11,50	
....	3.00	5.22	2.22	104.73	16,50	Side of Road.
3.75	5.22	1.47	....	108.48	17,00	Centre of Road, width 50.
7.44	9.28	1.84	....	115.92	18,00	Side of Road.
7.98	10.92	2.94	....	123.90	23,00	
....	3.53	9.07	5.54	118.36	26,00	Side of Road.
2.89	4.19	1.30	....	121.25	B.M.	On bottom hook of gate.
....	1.30	10.47	9.17	112.08	27,00	Centre of Road, width 40.
....	1.84	5.00	3.16	108.92	27,50	Side of Road.
....	5.00	11.77	6.77	102.15	30,00	
....	.58	11.07	10.49	91.66	35,00	
....	1.93	12.35	10.42	81.24	37,00	
....	1.19	11.52	10.33	70.91	40,00	
....	1.08	9.10	8.02	62.89	46,00	
5.10	8.90	3.80	....	67.99	54,00	
....	3.80	11.70	7.90	60.09	57,00	
....	1.85	11.12	9.27	50.82	61,00	
....	2.78	8.20	5.42	45.40	64,00	
....	8.20	9.12	.92	44.48	70,00	
3.57	5.85	2.28	....	48.05	79,00	Side of Road.
.05	5.05	5.00	....	48.10	80,00	2 Miles. Centre of Road at——
1.50	5.00	3.50	....	49.60	,50	Side of Road—width 100.
....	3.50	11.66	8.16	41.44	6,00	
....	2.90	8.10	5.20	36.24	10,00	
....	8.10	10.90	2.80	33.44	14,00	
....	1.78	7.15	5.37	28.07	24,00	Occupation Road, level with adjoining fields.
2.01	7.15	5.14	....	30.08	30,00	
9.00	9.50	.50	....	39.08	38,00	
5.82	8.00	2.18	....	44.90	45,00	47.50. Occupation Road, level with adjoining fields.
.29	6.45	6.16	....	45.19	57,00	
1.41	7.95	6.54	....	46.60	B.M.	Bottom hook of gate.
2.54	6.54	4.00	....	49.14	63,00	
....	4.00	8.92	4.92	44.22	64,00	Centre of Lane, width 30.
5.50	8.92	3.42	....	49.72	65,00	
4.43	8.27	3.84	....	54.15	73,00	
8.24	10.37	2.13	....	62.39	79,00	
124.8	289.80	355.79	190.87			
		289.80	124.88	128.38	....	Red. level at top of page from which last reduced level is deducted.
		65.99	65.99	65.99		

Elevation.	Back Sight.	Fore Sight.	Depression.	Total Rise.	Distance.	REMARKS.
				62.39	.....	
....	8.68	12.14	3.46	58.93	85,00	3 Miles.
	2.13	11.73	9.60	49.33	9.50	In road, side of park.
5.70	8.90	3.20	......	55.03	B.M.	Bottom hook of lodge gate.
2.20	3.20	1.00	......	57.23	16,00	
......	1.00	3.45	2.45	54.78	20,00	
......	5.38	9.47	4.09	50.69	29,00	
......	3.08	5.20	2.12	48.57	35.00	In Lane, level with fields.
.....	5.20	5.87	.67	47.90	42,00	
7.36	9.14	1.78	.....	55.26	46,00	
......	1.78	7.65	5.87	49.39	52,00	
......	2.10	11.85	9.75	39.64	56,00	
......	1.24	10.75	9.51	30.13	B.M.	Bottom hook of gate.
15.26	51.83	84.09	47.52			
		51.83	15.26	62.39	......	Red. level at top of page, from which last reduced level is deducted.
		32.26	32.26	32.26		

The method of procedure in setting down, casting out, and reducing the levels in the above form of field book, will be easily understood by a slight inspection. In the second and third columns opposite to each other, are entered the back and fore sights, the first column containing the difference of the two if a rise; if a fall it is entered in the fourth,* the differences, it is evident, by this form, being set down in the order they are cast out. The fifth column contains the reduced levels, which are obtained by adding or subtracting the differences to or from the previous reduced level; the sixth column contains the distance, which, it will be perceived, is continued from the commencement to the termination of the section; and the last column, which is the largest, is reserved for remarks on bench marks, crossings of roads, brooks, rivers, &c., as well as the geological features of the country; and, if

* When the differences of columns 2 and 3 have been cast out for a page, the four columns should always be added up at the bottom, previous to reducing or carrying the difference to column 5, and the sum of the differences will, in every instance correspond, if correctly computed. By deducting the sum brought forward at each page from the last reduced level the remainder will also correspond with the sum of the differences if the whole is correct.

necessary, to enter the bearing by compass of the sectionnl line; in the above case, no bearings were necessary from the line being well defined. In the forms of field book generally used, there is a separate column for the bearing at each station, but which is quite useless; it is sometimes indeed necessary in running trial levels through an unenclosed or thickly wooded country, where there is danger of swerving from the correct line, to note the bearing of an object for the chainman to run to, and as often as the direction of the line is changed as often to note the bearing. By such means the sectional line may be subsequently laid down on a map of the country, when it can be seen if any improper deflections have been made; but it will not be necessary in any case to note the bearing oftener than the *general* direction of the line is changed.

The section represented at plate 8 is plotted from the above field book, to a scale of 20 chains to the inch horizontal (4 inches to the mile), and 50 feet to the inch vertical; the roads, &c., written on the section, refer to that tinted brown; if similar particulars of those tinted red and blue had been considered necessary, such would have been written with the same colours as the respective sections are tinted with. The distances in miles, and also the vertical and horizontal scales, should always be drawn and written on a section; and many engineers require all the horizontal distances and vertical heights to be figured, as a check on the accuracy of the plotting, which also enables any other person to replot it, if required to a different scale.

At crossing all roads, lanes, brooks, rivers, &c., the local names should be ascertained and written on the section, with the extreme height to which floods rise wherever such occur; also the names of places to and from which the intersected roads lead;—in fact, *too much information cannot be obtained*, for particulars which at the time often appear trifling, afterwards become of the greatest moment. It would also be a matter of considerable importance, to note the description, quality, and probable value of land along the several lines, and whether consisting of clay, gravel, chalk, &c. &c., the price of building materials, labour, facilities of land and water carriage, &c., as on such particulars the eligibility of a particular line mostly depends.

The advantages of reducing levels from an assumed datum line fixed at a lower level than the presumed lowest point on the section will be very apparent if the reader will take the trouble to reduce and plot the levels in the preceding example from the level of the first station, of course taking that as 0.00. He will then perceive the trouble and danger of having *minus* quantities to plot, although where such do unavoidably occur—as in levelling over fen districts where the Trinity high-water mark is taken as the datum,—a column is generally set apart in the field-book, and headed, "total depression." In some few cases this plan of field book may be adoptod with advantage; as when several persons are employed in taking different portions of a continuous section, each might then reduce and plot his levels from a mutual bench mark at the exremity of the division. The difference of level of the termini might be afterwards easily computed by bringing forward the difference of the several mutual bench marks, and the plotted section if connected by the surface line, should always be checked by the computed difference of the terminal points of the several portions of the work.

We have not yet hitherto mentioned the subject of "intermediate sights" in levelling, but as a clear understanding of their use is of the utmost importance, not only in economising time in the field, but also in enabling the operator to represent the most minute and trivial deflections in the ground line, we shall devote a short space to the subject. From our previous remarks on the method of levelling by back and fore sight, it might be supposed by some of our readers that *two* sights only can be taken by once planting the spirit-level—*i. e.* a back and a fore sight; but nothing is more erroneous, for any number of sights may be taken from one instrumental station with as much accuracy as two sights. We can at once elucidate the matter by reference to the following diagram:—

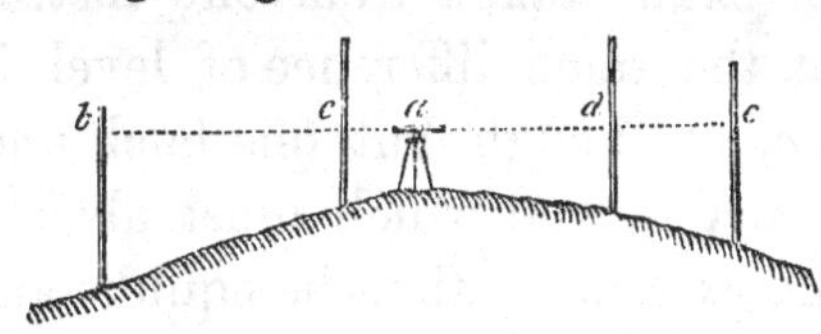

where *a* is the spirit-level with the staves *b* and *c* planted equidistant therefrom. Now, suppose the reading at *b* to be 10 feet, and at *c* 8 feet, the true difference of level between those points would be 2 feet; and if while the level was directed towards *b*, a staff was set up at *e* the reading on which was 4 feet, the difference of level between *b* and *e* leaving curvature, &c. out of the question, would in such case be *plus* 6 feet. But if we were still further to forego the slight theoretical errors arising from unequal distances, and compare the readings at *e* and *c*, we should find the difference to be *minus* 4 feet; which deducted from the previous *plus* difference, would give 2 feet from *b* to *c* as before. In a similar manner might we take another sight at *d*, or any number of sights *within b* and *c*, without vitiating the result of the levelling, by curvature, refraction, imperfections of the instrument, &c. for the errors arising from all such sources would be wholly confined to the intermediate sights *c* and *d*, which the reader will at once perceive by throwing the readings into a tabular form.

Rise.	Back Sight.	Fore Sight.	Fall.	Reduced Levels.	Distance.	REMARKS.
				20.00	......	Above assumed datum.
6.00	10.00	4.00	......	26.00	......	Reading at *e*.
......	......	5.50	1.50	24.50	......	.. „ .. „ *d*.
......	......	8.00	2.50	22.00	......	.. „ .. „ *c*.
6.00	10.00	8.00	4.00	20.00	Deduct	from assumed datum.
4.00	8.00					
2.00	2.00			2.00		

In this example it will be perceived that there are three fore sights and one back sight taken from one instrumental station, and that the same difference of level is obtained between *b* and *c*, as though but one back and one fore sight had been taken, and which must always be the case whether the extreme stations be equidistant from the spirit-level or not. For if we consider that in the readings of the staves at *e* and *b*, there will be a certain amount of error by the instrument being so placed, we shall also perceive that a reverse error will take place in the readings of the staves at *e* and *d;* which although it may not be sufficient to neutralise the preceding error, will still tend that way; but on the readings of the staves *d* and *c*, there must necessarily be an amount of error, which, together with that accruing from the preceding sight, will equal that which arose in taking the observations at *b* and *e*.

It is the practice of many engineers to register intermediate sights in the same manner as we have in the preceding example,* while others have separate columns for their intermediate field-registry, as also for their

* When such is the case the computation of levels is proved by adding up all the back sights, but only the *final* foresights, and deducting the one from the other as by the former method.

resultant levels, thereby greatly increasing the complexity and bulk of the field book. By reference to the field book at p. 40, and opposite to the B. M., immediately past three miles in the distance column, there will be found three fore sights taken from one instrumental station, the registry of which is somewhat different from, but far preferable to, either of the preceding methods. The first fore sight is 3.20, which is entered as a back sight to the next forward observation of 1.00, which quantity is again entered as a back sight to the next fore sight of 3.45. By this form of registration there is a repetition of each intermediate fore sight entry, but which not only simplifies the computation and affords a more ready means of testing its correctness, but enables the operator to cast out the differences of level in the field during his temporary stoppages for the chain or staffmen, thereby saving much time in the office, as little more would then remain to be done but plot the section. The greatest care is, however, necessary in entering an intermediate sight in the column of back sights, as if any discrepancy occurs between the two entries, the amount of difference will be an error which will continue through the whole of the subsequent levels; for this reason we feel inclined to recommend that ordinarily a final fore sight should close each page. A similar caution is also requisite as to the transfer of the reduced levels from page to page of the field book, as we have sometimes known most serious errors to be introduced solely from carelessness in these particulars; we would therefore advise all persons engaged in levelling operations, to cast out and reduce their own levels, even though they have ever so many assistants, for the most careful observer may have his reputation destroyed, and will always in fact be in danger, if he commits to others the responsible but certainly laborious task of reducing an extensive series of levels; but by casting out each day's observations as the work proceeds, little trouble will be experienced in the reduction. Respecting the method of registration which we give the preference to, we have merely to remark that we have found it adopted by the most expert and practical men with whom we have come in contact—a sufficient guarantee we think, of its superiority.

We have previously remarked, that in levelling operations there is no check on the accuracy of the performance except by going over it a second time; and which is in fact invariably the case where the levels are of any importance; the second operation is termed "Check Levelling," but is very differently performed from the first. At various points along the line are left bench marks from the trial levels (generally at about every two miles) which if not previously determined on, are very minutely described by the person leaving them; and to render their recognition more certain, a distinguishing mark should be made and noted in the field book for the guidance of the person coming after. To be certain that no errors have been committed in the original levels, it is usual for a second person to check them, by ascertaining the difference of level between the various bench marks; when if the differences prove to be the same as before, or nearly so, it is fairly to be inferred that the intermediate levels forming the section are also correct. The person taking the check levels does not generally pass over the sectional line, but proceeds by the nearest and most convenient route from bench mark to bench mark,—generally by a road if running near, without noticing the variation of the ground, and taking as moderate long sights as his instrument and the ground will permit. No chain is required in this kind of levelling, but the observer should be careful to plant his staff (as near as his eye will direct him) at equal distances on each side of his instrument; errors arising from curvature, imperfections of instrument, &c. will then be rendered nugatory. As an example of this kind of levelling, we may refer to plate 8, where part of a trial section tinted brown was decided on as the best; but there not being time to level it in detail, being required for immediate deposit with the Clerk of the Peace (a practice never to be resorted to if by any possibility it can be avoided), it was rapidly check-levelled, for whatever it wanted in detail, it was necessary to have the general results correct. For this purpose levels were taken to various bench marks along the line—it would have been a waste of time to have taken all that were left, but only to those that were conveniently situated, or of material consequence, *as the crossing of a summit, or any particular road or point on the line to which reference might be made by an opponent.* The first bench mark on the line that was levelled to was at the stile, which was really unnecessary, being so short a distance from the commencement, but as it laid very convenient it was taken. We append the check levels up to that point, which will fully explain the method :—

Field Book.—Check Levels from B. M. on Root of Tree, assumed 100 *feet above Trinity High-water.*

Elevation.	Back Sight.	Fore Sight.	Depression.	Total Rise.	Distance	REMARKS.
	......	......	......	100.00	......	B. M. above Datum.
	8.60	2.20				
	9.45	3.20				
	5.40	18.0				
	4.51	9.27				
	4.86	8.25				
	13.40	10.52				
	5.60	6.90				
	5.80	9.43				
	8.83	5.64	......	109.24	......	B. M. on Stile.
	66.45	57.21				
	57.21					
	9.24	Difference of Bench Marks				

It will be seen on reference to the former field book, in which the height of the B. M. at the stile is entered, that the difference of the two levellings was but $\frac{1}{100}$th of a foot; it was therefore assumed that the intermediate levels were correct. In a similar manner were the check levels taken along the whole line; and in one or two instances where bench marks had not been left, the level of the ground was taken, as at the crossing of a road, and at the summits. The results of check-levelling may be obtained by casting out the differences of each back and fore sight—no intermediate sights being taken, and adding up the *four* columns at each page to prove the computation; but of course no general reduction of the levels is requisite. But so much trouble as this is rarely taken, the differences being generally obtained by casting up the columns of back and fore sight, and deducting the one from the other, checking the computation by casting the columns a second time, commencing from the top. We have in this example given the same form of field book as when levelling for sections, although it is evident but two columns are requisite; but it is presumed that the field book is ruled and printed throughout in the same manner.

We now proceed to give a practical illustration of levelling, in which greater detail is observed than in the example which we have given of a trial section; indeed, under ordinary circumstances, the minute particulars embraced in the accompanying railway section would be sufficient for the engineer to estimate the cost of construction and lay out the works. We intended to have given some examples of working sections and level books in this portion of our work, but think it better to retain such for that part which treats of the "Laying out of Works."

The section represented at Plate 7, is plotted from the accompanying field-book to a scale of 5 chains or 330 feet to the inch horizontal, and 40 feet to the inch vertical. The field-book is on precisely the same system as the former example; but additional information it will be perceived is comprised in the last column, where the crossings of the several roads, &c., are sketched, with their dimensions; which, being plotted to a large scale on the section, are found of infinite service to the engineer in guiding him as to the dimensions of the bridges, so as to maintain a proper width of road or water-way, at least so far as existing circumstances,—which will generally be found the surest guide,—can point out.*

Field-Book.—" Distances measured with a 100 *foot Chain."*

Elevation.	Back sight.	Fore sight.	Depression.	Total Elevation.	Distance.	REMARKS.
....	....	....	....	80.96	B.M.	Above Trinity high water.
1.11	6.84	5.73	....	82.07	0,00	Level of ground at commencement.
....	5.73	8.10	2.37	79.70	,15	
....	8.10	8.15	.05	79.65	,45	
2.35	8.15	5.80	....	82.00	,50	
.80	5.80	5.00	....	82.80	1,10	
.45	5.00	4.55	....	83.25	2,17	
.96	5.01	4.05	....	84.21	2,60	
....	4.05	4.98	.93	83.28	4,00	
....	4.98	6.12	1.14	82.14	5,00	
....	6.12	6.67	.55	81.59	6,00	
....	2.25	7.77	5.52	76.07	8,00	
....	7.77	13.52	5.75	70.32	10,00	
....	3.95	6.30	2.35	67.97	11,00	
....	6.30	10.80	4.50	63.47	12,00	
....	1.60	4.24	2.64	60.83	13,00	
....	4.24	6.74	2.50	58.33	14.00	
....	6.74	10.20	3.46	54.87	16,00	
....	2.17	4.60	2.43	52.44	18,00	
....	4.60	5.36	.76	51.68	20,00	
....	5.36	5.99	.63	51.05	21,00	
5.67	106.76	134.67 104.76	35.58 5.67	.... 80.96	....	Level at top of page from which last reduced level is deducted.
		29.91	29.91			

* From want of attention to "existing circumstances," the London and Birmingham Railway Company after an expensive legal process, have been compelled to reconstruct several of their bridges so as to preserve the original width of roadway, notwithstanding that they were in the first instance built of the full width particularised in their act of parliament.

Elevation.	Back sight.	Fore sight.	Depression.	Total Elevation.	Distance.	REMARKS.
				51.05		
....	4.49	5.62	1.13	49.92	23,00	
....	5.62	7.57	1.95	47.97	25,00	
....	3.16	4.58	1.42	46.55	27,00	
....	4.58	4.92	.34	46.21	28,00	
....	4.92	5 20	.28	45.93	29,00	
....	5.20	6.19	.99	44.94	30,00	
....	4.02	5.21	1.19	43.75	31,30	
....	5.21	6.08	.87	42.88	31,57	Centre of Road.
.92	6.08	5.16	....	43.80	31,80	
....	5.16	9.27	4.11	39.69	33,50	
....	2.55	5.30	2.75	36.94	B. M.	On bottom hook of Gate.
....	5.30	6.15	.85	36.09	35,00	
....	6.15	8.73	2.58	33.51	36,00	
....	8.73	12.52	3.79	29.72	37,30	
.70	5.90	5.20	....	30.42	37,72	On bank of River.
....	5.20	9.60	4.40	26.02	37,83	Centre of River-bed.
4.66	9.60	4.94	....	30.68	37,92	Bank of ditto.
1.96	4.94	2.98	....	32.64	B. M.	
....	2.98	6.60	3.62	29.02	38,60	
....	6.60	6.61	.01	29 01	40,00	
.85	8.70	7.85	....	29.86	41,00	
2.33	7.85	5.52	....	32.19	42,00	
1.98	5.52	3.54	....	34.17	43,00	
....	7.29	8.10	.81	33.36	43,25	Side of Road.
....	8.10	8.20	.10	33.26	43,50	Centre of ditto.
4.33	8.20	3.87	....	37.59	B. M.	Side of do.
....	3.87	4.78	.91	36.68	45,00	Set off on Wall.
1.23	4.78	3.55	....	37.91	46,00	
1.28	3.55	2.27	....	39.19	47,00	
.72	6.72	6.00	....	39.91	48,00	
.24	6.00	5.76	....	40.15	49,10	
1.60	5.76	4.16	....	41.75	51,00	
1.21	7.36	6.15	....	42.96	53,00	
1.25	6.15	4.90	....	44.21	53,80	
....	4.90	6.28	1.38	42.83	53,97	Centre of Chase.
.08	6.28	6.20	....	42.91	55,00	
.45	6.20	5.75	....	43.36	56,00	
....	4.70	4.72	.02	43.34	59,00	
.83	4.72	3.89	....	44.17	61,00	
.48	5.81	5.33	....	44.65	63,00	
.83	5.33	4.50	....	45.48	66,00	
.08	4.50	4.42	....	45.56	67,00	
.55	6.00	5.45	....	46.11	69,00	
.12	5.45	5.33	....	46.23	71,00	
....	5 33	5.71	.38	45.85	73,00	
....	3.30	4.75	1.45	44.40	74,60	
....	4.75	6.12	1.37	43.03	75,50	
....	6.12	7.09	.97	42.06	77,00	
....	4.23	4.94	.71	41.35	81,00	
....	4.55	5.18	.63	40.72	83,00	
....	4.74	5.14	.40	40.32	85,60	
.36	5.14	4.78	....	40.68	85,95	Centre of Road.
....	4.78	5.10	.32	40.36	B. M.	
29.04	993.07	303.76	39.73	....		Level at top of page, from w hich last reduced level is deducted.
		293.07	29.04	51.05	....	
		10.69	10.69	10.69		

If the reader should not now fully comprehend the particulars registered in the field-book, we should recommend him carefully to examine it in connection with the section; to go through the casting out and reduction of levels, and then to plot the section from the field-book, which cannot fail in removing every difficulty from his mind.

Before proceeding further with our subject we have a few general remarks to make on levelling operations which the present appears a fit place for introducing. Curvature is rarely allowed for in practice, as although the spirit level may not always be planted midway between the staves, yet the mean of a great number of observations will give the correct *practical* difference of level between any points, provided the inclination of the ground is not continuous. But even so, if the operation of levelling is properly conducted, the errors from this source will in most cases be too trivial to be noticed. In passing over an undulating line of country of course what is gained from curvature, &c. in an ascent, will be lost in the next descent, so that practically speaking the results would be correct; for the method of allowing for curvature, the reader is referred to the introductory chapter. But it should be observed, that it is *always desirable* in levelling operations to place the staff as near equal distances (roughly estimating it with the eye) on each side of the spirit-level as can be done conveniently; which, if then out of adjustment, will neutralise the errors which would otherwise be occasioned. But if the adjustments are perfect, it is not necessary that the staff should be so placed to ensure accurate results, in practice the operator generally choosing the most commanding spot of ground on which to plant his instrument, without regard to its being exactly in a line with, or midway between, the staves; although, as we have before observed, it should be attended to when it can be conveniently.

We may here remark that we consider it preferable to use one staff in place of two, as is common with many persons, on account of differences which occur in the graduations, although made by the same person with ever so much care. We have frequently, when adjusting our instrument, found a difference of $\frac{1}{100}$th of a foot in the readings of two staves, even when made by the same person; and no doubt the errors are often much greater. To avoid this source of error, we would always advise the use of one staff in preference to two, notwithstanding many persons may say that the two staves becoming alternate back and fore sights, will neutralize any error that may be supposed to arise from a difference in their graduations. For in a rugged district where such difference would be called forth at each planting of the instrument, it would occasion great trouble, and be almost impossible for the operator always to observe that the staves were placed *alternately;* and without doing so a permanent error would be introduced in the levels. The errors which would arise from such source would undoubtedly be small, but possibly by frequent repetition might become a sensible quantity, although the little time saved by the use of two staves in taking preliminary sections might be of more consequence than extreme accuracy in the levels; but in such case there must be an extra expense incurred, greater damage to crops, and consequently rendering opposing parties more hostile; and lastly, we confidently assert, that the chances of error are greatly multiplied. We would therefore earnestly recommend that in taking contract sections, or in laying out works where a slight saving of time cannot be of such consequence as rigid accuracy, that one staff only be used; and let the observer, even in such case if the staff is jointed, be careful to prove its accuracy when put together, by measuring a foot at each joint, taking it half above and half below, which of course should be of exactly the same length as a foot on any other part of the staff not intercepted by a joint.

By the use of one staff, however, in *preliminary* sections,—where of necessity the operator will procure men unaccustomed to the duties, and in most instances indifferent and careless as to the results,—some difficulty will be experienced in the alternation; for, if after observing the last fore sight, and before re-adjusting the instrument and obtaining the next back sight, the staff-holder should move from his position with the idea of being in readiness for the next fore sight (by no means an uncommon occurrence where dispatch is requisite), all means of reference to the preceding levels will be lost. The operator will in such case have the choice of *guessing* at the previous position of the staff and taking its level to enable him to continue onward with the section, or otherwise retrace his steps to the last bench mark, and recommence levelling therefrom. Particular instructions should however be given to the assistants on no account to stir until *positively* signaled by the principal so to do; and at each setting up of the staff the holder should be directed to press it gently on the ground, so that the weight of the staff will have no influence in causing a change of level, but on *reversing* it for the back sight, to do so with all possible tenderness.*

In our examples of levelling it will be perceived that the chainage, or distance, is continuous; which, although it appears the most simple, and is certainly the most cor-

* For further remarks on this subject, see description of levelling staves.

rect method, is by no means generally adopted. It is common to enter the distance from staff to staff, or staff to instrument, in a column set apart for that purpose in the field-book, then add them together and enter the *total* distance in another column; but there are serious objections to such a method, as there must necessarily occur small fractional parts in the separate distances, which, if noticed, would be very troublesome, and probably introduce errors; and, if not noticed, would certainly do so. There will also be great confusion and liability to error from the irregular changing of the pins, and from many other causes which we think should prevent any person from pursuing such a system. But probably a worse custom than the preceding, is, in allowing the assistant to enter the distances in a book kept by him for that purpose, which distances are not entered in the principal's book, but referred to by letters or numbers; for if by any error or omission on the part of the principal or assistant in placing a wrong letter or number to the distance, the whole must be involved in doubt, and become worthless. We have known instances of ruinous errors being committed entirely from this method of registration, and we would urge those persons who are in the habit of keeping their field-books in such manner to abandon it at once for a simpler and safer plan.

We have before remarked that a section should be commenced from, and where possible ended at a bench mark; intermediate bench marks should also be left at various conspicuous and convenient points along the line, either for checking, varying, or continuing the levels at a subsequent period. In the example of levelling at page 50, the levels were commenced from a bench mark formed by a notch or bench cut on the root of an oak tree near to the line, and which was previously ascertained to be 80.96 feet above Trinity datum. The staff being first placed on the bench mark, and then on the ground at the point of commencement, gave a difference of level of 1.11 feet, which being a rise, was added to the height of the bench mark, thereby making the point of commencement on the section 82.07 above Trinity datum. The levels were then carried on in the usual manner to the crossing of the first road, where a bench mark was left for the purpose of subsequently taking the levels of the roadway for a distance on each side of the line to determine on the best mode of crossing, either by a bridge over or under it, and the quantity of excavation or embankment which would be required in forming the necessary approaches. Also the quantity of land (if any) which it would be necessary to purchase in forming the approaches, beyond the original site of road, and to determine on the best method to be pursued for effectually draining both the rail and roadway.

The method of arriving at the levels of the bench marks just alluded to, is, to set down the reading of a staff placed on any one of them as a forward station; but on placing the staff at the forward station on the ground, the reading on the bench mark becomes the back station to it; this will at once be understood by reference to our registry of levels. But where a section is finished on a bench mark, the reading becomes the last forward station; the reading of the staff placed on the ground at the extremity of the line of section becoming the back station to it. Of course when the levels are renewed, this bench mark becomes the back station, the same as at the commencement of the section in our example. At the close of a day's work, if a bench mark cannot be found convenient, drive a peg into the ground until its top is firmly level with the surface, on which place the staff, the reading will be the last foresight, and also answer for a temporary bench mark. On resuming the levels the next morning, the operations will be carried on as though no suspension had taken place; not recording this temporary bench mark in the field-book, but the distance in the proper column as usual.

In concluding this chapter, we may with advantage refer to the necessity of extreme caution in registering the observations at the several back and fore sights. We observe the following plan:—On first examination of staff, enter the feet and first place of decimals, referring to it a second time for the second place, and a third time to be certain the entire entry corresponds with the observation. When taking a back observation, if the spirit-level should be so near to the staff, that the range of sight does not embrace a *registered* foot, but the bisection be somewhere midway, it will be found desirable after entering the decimals, for the holder to incline the staff but without moving the foot of it, from, or towards the observer, until he can distinctly observe the feet thereon. For an intermediate or final foresight under similar circumstances, it would be sufficient for the holder to raise the staff above the ground to enable the observer to register the feet, the decimals being registered in the first instance. This plan is preferable to the ordinary practice in such cases, of looking along the outside of the telescope tube for roughly determining the level; or otherwise by making the holder place his hand or some object at the point of intersection, whereby the *feet* is subsequently read off by the observer after determining the decimals.

CHAPTER IV.

ASCENDING AND DESCENDING LEVELS.—LONG AND SHORT SIGHTS.—OBSERVING AND REGISTERING.—REDUCTION FROM INTERMEDIATE BENCH-MARKS.—INVERTED LEVELLING, REGISTRATION, AND RETROGRADE LEVELS.—EXPLORE AND FLYING LEVELS.—REDUCTION OF DISTANCE.—SECTION PAPER.—GENERAL REMARKS.

IN the last chapter we pointed out the method by which the effects of curvature and refraction are, in practice, neutralised, but such method it is clear will only be strictly correct when the elevations and depressions are equal. Where a section is taken, from the sea for instance, inland over a tract of country, which, although it may undulate, still gradually rises with the distance, it will be found that there will constantly be a preponderance of long back sights, by which means a certain error—small, it is true,—will be introduced, without a proper correction is made for curvature and refraction, on the excess in length of the back over the fore sights. This error will be *plus* the actual elevation, from the staff being apparently depressed by the curvature, and of course the bisection made at a greater elevation thereon. On the contrary, in taking a descending section—as from a summit to the sea, there will be a preponderance of long fore sights; in this case the error will be a *minus* quantity, and, of course, add to the depression. The difference from the true elevation or depression it is evident will be the same in both cases; but although neither operation will be correct, still the difference of level will be the same if accurately performed in other respects. Although we have thought proper to make these observations, we by no means consider it essential in the ordinary operations of levelling, to make corrections from the observed differences of level either in an ascending or descending section. Notwithstanding, in matters of experiment or scientific inquiry, in determining the level of a *standard* trigonometrical point, or other similar object, it would be necessary to make the correction. It should not fail to be observed that in taking a section—say from London to Brighton,—the termini would appear, or be at the same level, or very nearly so, if brought down to the mean of the tides at either place; therefore, what was gained in ascending the summits between those places would be lost in their descent; the summits would consequently appear slightly higher than in reality was the case, and points at the same level on opposite sides of the summits might fairly be supposed to have the same amount of error attached to them, which would, of course, become less and less as the elevation decreased.

We cannot quit this subject without allusion to a very erroneous notion which we have found to prevail with many persons in regard to an attempted correction. The idea has been this—by removing the spirit-level a short distance out of the *direct* line of section, it was supposed more equable sights were obtained. A second thought would have shown the fallaciousness of the first; but, unfortunately, too many of us are prone to proceed on first ideas, or on whatever we have heard, or seen practised, without troubling to inquire into the truth of the matter. So far from the above notion being correct, it will be found after a little consideration, that when the spirit-level is planted on the *direct* line of section, the staves being set up as far on each side of the instrument as the ground will permit, that then the sights approximate nearer to equality than can by possibility be the case in any other position of the instrument. But we do not mean equality of distance, but equality of errors, for, as the curvature increases as the square of the distance, any plan which increases the length of the sights, must increase the curvature in a much greater ratio than the approximation in the length of sights can counteract.

In all delicate operations of levelling over a quickly ascending or descending line of country, it would be desirable to ensure uniformity of sights by using a staff of exactly twice the height at which the spirit-level is usually set up above the ground, or otherwise only to bisect the *lower* staff at a less elevation than twice that of the level. By such means equal sights and accurate observations would be obtained, for the ground—as we have supposed, rising at the same rate, the *upper* staff would be bisected at the same quantity above the ground, as at the lower station it was below twice the elevation of the instrument;—thus, the spirit-level being set up 5 feet above the ground bisects the lower staff at 9.50 feet at any distance; now, if the ground continues at the same rate of inclination, it is evident the bisection of the staff at the upper station will be .05 at the same distance; in the latter case, being .05 above the ground line, in the former .05 below twice the elevation of the instrument.

These observations bear forcibly on the subject of long or short sights, the relative advantages of which we shall endeavour to place before our readers in as few words as possible. It will be at once perceived that, in taking a section between any two points, the less the number of sights in which it may be taken, the greater will be the distance of the staves apart; consequently, less time will be required in the operation. By such means, however, the greatest accuracy will not be secured, from the difficulty of approximating to equality in the back and fore sights, without which very sensible errors will be introduced. In sights of 3 or 4 chains, or 5 chains, it signifies little whether a back sight is a chain longer or shorter than its fore sight; the error will be quite inappreciable, providing the instrument is in adjustment. But the case is widely different when we come to sights of from 10 to 20 chains, where it will be necessary that the back and fore sights very closely approximate indeed. We will draw out the results for the reader's information. By the table at the end of our work it will be seen that—

	Feet.
The correction for curvature and refraction is for 5 chains	.00224
Do. do. . . . . . . for 6 chains . .	.00321
Error being . .	.00097
Correction for curvature and refraction for 19 chains	.03225
Do. do. . . . . . for 20 chains . .	.03573
Error being . .	.00348
Deduct former error	.00097
Excess of error in the latter case	.00251

Many of our readers will doubtless say, even by our own showing we have not made out an appreciable error, seeing that the field registry is not carried beyond two places of decimals. We may observe, in reply, that in very long sights it will rarely indeed occur for the back and fore sight to approximate so closely, setting aside the difficulty of the observers enforcing attention to that point. We may fairly assume as a moderate allowance for deviation in length of sights carried to such an extent, that the difference would be in the same proportion as the first example. It would then stand thus:—

	Feet.
Correction for curvature and refraction for 20 chains ,	.03573
Do. do. . . . . . for 24 chains . .	.05146
Error on excess of 4 chains	.01573

In this case there would be an appreciable quantity (. 02), which, although small, would soon increase by repetition; and which, coupled with many other little sources of almost unavoidable error,* would render a section taken under such circumstances an unfaithful, and perhaps dangerous document. On the other hand, if short sights are insisted on, much more time is consumed in the operation, and from the more frequent transference of level by the greater number of sights, errors will creep in without the greatest care and circumspection is observed. Where a long line of levels is to be taken by a system of short sights, it will be absolutely necessary to introduce a third place of decimals to insure accuracy, in consequence of the frequent *bisection* of the hundredth divisions making it doubtful with the observer whether to give or take the .005: if the former in one case, he should take it on the next recurrence to balance the error. If he should neglect to do so, an error is involved; if he should take it on the wrong side it is doubled, and becomes a sensible quantity. The more preferable method is, as we have suggested, to preserve the same graduations, but subdivide the foot into thousandths by *estimation*, which so far as $\frac{5}{1000}$ths, or $\frac{25}{10000}$ths are concerned, can be done with considerable

* The mechanical defects of instruments will in such cases be brought prominently forward, and with the many little disarrangements which every instrument is liable to, from remaining so long in one position, from the action of the wind, reversing, &c., will all tend to increase the error.

accuracy. As a general rule, we should, however, say give the .005 on the longest sight, take it on the shortest, curvature will explain the reason why. Perhaps some of our readers may prefer the following method of clearing the readings of such minute parts without vitiating the whole operation. On bisecting the $\frac{1}{100}$th of a foot with the *last* fore sight, desire the staff holder to strike the staff down smartly until the bisection is exactly *at* the $\frac{1}{100}$th. When the *back* sight is similarly circumstanced, the desired bisection may be made by a slight pressure on the level legs. This is another case in which many people will come out with their "average," which they may say will produce a mean of error, and consequently accurate results; but this, as we have shown before, will depend on chance, and that a very doubtful one; but surely it must be better in all cases to investigate and allow for errors than trust to chance to correct them.*

The act of observing levels may appear too familiar a subject to enlarge on; but under that head we embrace the position and planting of the level and staff, in addition to the registration. The position of the spirit level, in regard to the staves, must in all cases be determined by the nature of the ground and the degree of care required in the operation. Except under peculiar circumstances, the most accurate and advantageous position in which to place it is, on the *direct* line of section, midway between the staves, and on ground of *medium* elevation. The advantages of placing the level on the direct line of

* If the reader should be desirous of ascertaining by experiment the most correct method of obtaining accurate results with his *individual instrument*, let him fix on a bench mark and level a distance out and back, first by long and afterwards by short sights. By such means, the difference, if any, which may occur in either series of levels, will be an error; and whichever may be in excess will be the most inaccurate method, all circumstances being the same.

section is not confined to where the ground is of quick ascent or descent, but is so under any and every circumstance. We have said that ground of *medium* elevation should be chosen on which to plant the spirit-level; for by so doing a difference of level may be taken at one sight equal to the difference between the whole height of the staff and that of the instrument. We should not, however, be understood by the preceding expression always to recommend etheir medium or the highest ground; by reference to the annexed diagram,

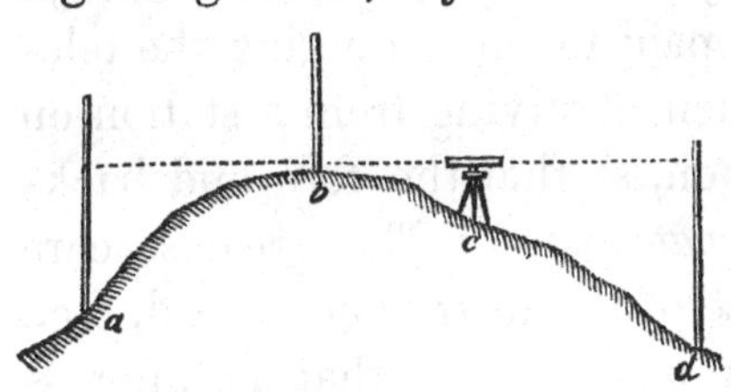

it will be seen that it would be a disadvantage to select either one or the other for such purpose; at the same time it will show that time may be greatly economised by attention to many little matters either unnoticed or considered too trifling to be attended to by ordinary observers. In passing over a summit, or ridge, *a, b, c, d,* instead of planting the spirit level at *b,* the highest point, set it up at *c*; *superior* to the back station *a,* as well as to the forward station *d.* By such an arrangement the staves are bisected at their very summits; consequently the greatest possible difference of level is obtained in one sight. Again, by placing the spirit-level as in the diagram, the line of sight just skimming the summit, a greater difference of level is obtained at one sight by the whole height of the instrument than if it had been set up at *b.*

On the actual planting and adjusting the instrument for observation, we have, for the present, but a very few remarks to make. In planting the spirit-level on the *direct* line of section, it is desirable in observing, for the telescope to lie over two of the parallel plate screws, for the purpose of being able to correct any *slight* alteration of level—consequent on reversing the telescope;* action of the wind, subsidence, accidental contact of the observer with the stand, or other cause—by motion of one screw without danger of *laterally* disarranging the level, as would likely be the case if such correction was attempted with the telescope in any other position. The danger of bisecting the level staff with any part but the centre of horizontal wire will be evident, from the possibility of the instrument being slightly out of level transversely. Attention should also be paid to the arranging the telescope over the screws when observing from a station on one side the line of section, so that the fore and back-sights shall lie over *alternate* pairs. The greatest care should be paid to the state of the instrument, stand, &c., and a daily examination made to see that nothing is amiss; all the screws should be carefully tightened up before commencing operations, as we have known errors to arise from a single screw in one of the legs being slack. In moderate windy weather considerable attention is requisite in setting up the instrument firm, and so prevent vibration as much as possible; one leg of the stand placed opposite to the wind's direction will be found to yield the most steadiness and safety. Where the line of section rises or falls quickly, or moderate long-sights are taken, much time will often be saved by *roughly* levelling the instrument and bisecting the back-staff to see if in range of level; when, if so, the final adjustment for observation may be made as usual, otherwise its position

* Instruments that will not reverse without derangement of level, are either not properly balanced, or the base of cone, or plate, or both, are untrue. When this is the case, it will be found of advantage to ascertain over *which* pair of screws the telescope will best reverse, and then arrange the instrument accordingly at each station.

must be altered. In this, as in most other matters, over-eagerness defeats itself. We have frequently observed persons remove their instruments two or three times for the purpose of obtaining the uttermost difference of level between their fore and back-sights, rather than secure a moderate difference in the first instance; in other cases, where they have been obliged to reset their instruments after adjusting, from having gone beyond range in the first instance—in both cases losing much time.

Of the registration of levels we have spoken in the last chapter, with the precautions we adopt to ensure accuracy. In addition thereto, when in doubt as to the correct division bisected on the staff, from constant changes while observing, take the lowest; for it will probably be the case that the holder is inclining the staff forwards or backwards, or the wind will so affect it if held ever so perpendicular; in either case, of course, causing the bisection to be made at a greater elevation than correct. This is remedied by taking the lowest reading, for it is evident that by no movement of the staff can a lower reading than the correct one be obtained.

In windy weather the cross wires of the telescope become so tremulous as generally to leave a doubt in the observer's mind as to the correct bisection. Such a circumstance must be carefully attended to before applying our last observations to practice, as it may so happen that the *least* reading will be inaccurate. For the vibrating of an instrument by the wind, it is but reasonable to suppose, would be as much above as below the horizontal line when the level is not *permanently* affected, consequently the *mean* reading would be correct if the staff was not also affected in some degree;—and this we have found to be the case in practice at very considerable distances, the reading being alternately as much above as below the correct bisection as ascertained during a temporary lull. But this has been, with the bisection, near to the bottom of the staff, and therefore not appreciably affected by its obliquity; during sudden gusts of wind the horizontality of the instrument will of course be affected, in which state no observation should be taken. These last directions should be paid especial attention to when reading the staff at considerable distances, for the angle of vibration being the same in long as short sights, of course the deviation from the correct bisection will be in proportion to the distance. Long sights should not however be resorted to in windy weather, or indeed at all, except under very auspicious circumstances; a dry, clear, calm day, with the instrument in perfect order and adjustment, will, in urgent cases, warrant sights of such length as would be reckless in any other position. When sights over or approaching 10 chains are taken, much depends on the accuracy of the observer in *estimating* the hundredths of feet; for although at that distance, with the best of level telescopes, he will have a very distinct perception of the tenth and perhaps of the twentieth part of a foot, he will be obliged to divide the latter with his eye, and then apportion what may appear to him the correct quantity bisected by the cross wire. Although this may appear a loose method of observation, still with a person having a keen eye for the perception of differences of magnitude, it admits of the greatest accuracy. A friend of ours, acknowledged to be a most correct observer, and whose levels always prove to the greatest nicety on being checked, never would use a staff graduated finer than the .025 of a foot, and he preferred it graduated only to .05, estimating a lesser quantity by his eye. This gentleman's practice, which has been most extensive, shows the inexpediency of the attempts which have been made to complicate the staff with contrivances for obtaining the bisection, even to the $\frac{1}{1000}$th part of a foot.

In case of *mirage*, or any violent ebullition of the atmosphere, in which the cross wires appear constantly to vary in their bisection, we have found the mean reading to be also correct, when the staff was quite perpendicular. In levelling towards evening great difficulty will be experienced in reading a staff west of the observer, but the contrary, when east; therefore, by giving a preponderance to the easterly sights, the operation, when urgent, might be continued for some time longer, than by taking equal sights as in ordinary cases. We have before observed that ruled level books with printed headings, should be used, and always of the same description, more particularly where several persons are employed on the same work. Care should be observed that the faint ruling is continuous in right lines across both pages, as if otherwise, through carelessness in the binding, wrong references may be made to an observation, and such we have known. On a levelling-book being filled with observations, it should be paged throughout, carefully indexed, and put by for reference, becoming a kind of stock in trade to an engineer, and that of the most valuable kind.

It will often happen in taking a detailed section after the determination of bench marks and the general direction of the line, that the observer is constrained to commence his operations at a point where no bench mark may have been left, and proceed to a considerable distance, or to the extremity of his section, before levelling on to one. In such case, if the distance is considerable, the inconvenience of adopting an assumed datum and reducing all the levels to it, is almost unavoidable; but where the distance is not greater than would take a few days to level over, it is a preferable method to cast out the work, but without reducing, and leave it in that state until the levels are taken up to the bench mark; the reduced level of which is to be inserted, and the subsequent part of the section proceeded with and reduced in the usual way. The method of reducing the previous part of the work is to *reverse* the usual method—adding *the falls* and deducting *the rises* from the posterior reduced level, as in the following example:—

Rise.	B. Sight	F. Sight.	Fall.	Reduced Level.	Distance in feet.	REMARKS.
				202.89	....	Level of Station at commencement.
1.50	3.81	2.31	....	204.39		
.80	5.31	4.51	....	205.19		
....	4.51	6.40	1.89	203.30		
....	3.38	6.68	3.30	200.00		
....	6.68	10.00	3.32	196.68		
2.18	10.00	7.82	....	198.86		
....	3.84	10.00	6.16	192.70		
7.66	10.00	2.34	....	200.36		
....	2.34	4.48	2.14	198.22		
....	4.48	7.12	2.64	195.58	....	Level of B. M.
12.14	54.35	61.66	19.45	......		
		54.35	12.14	202.89	....	Last reduced level.
		7.31	7.31	7.31		

It will be perceived that the last entry in the column of reduced levels is that of a bench mark previously known, and from which all the other reduced levels are worked out. Thus the fall 2.64 becomes a *rise* (as we are working) in connection with the *posterior* reduced level 195.58, and gives 198.22; the fall 2.14 is also a *rise* in connection with 198.22, and gives 200.36; again, the rise 7.66 becomes a *fall* in connection with 200.36, and gives 192.70; and so through the whole quantity. The difference between the first two sights, it will be perceived, is a rise of 1.50, but as it is a fall in connection with 204.39, it gives 202.89 as the reduced level of the bench mark or station at the commencement. If the bench mark, instead of being at one end of the line, as in the example above, should be in the centre, the same system would be pursued up to the bench mark, after which the levels would be reduced in the usual manner.

We have often seen old as well as young practitioners at fault in a case of this kind; where the bench mark was in the centre, working either way, and where at one end going there to commence.

Another point in connection with intermediate bench marks, where cross sections or *offsets* are registered with the main section is, to be careful in *re-inserting* their reduced levels when proceeding with the main section, provided a removal of the instrument is required. Thus, in the illustration of levelling with cross sections at plate 8, with the spirit-level at *a*, a sight can be just taken to *c*, where a cross section of the ground is necessary; but it will be perceived that the lower portion of the cross section *c e* can be taken without removing the instrument, but to take the upper part, *c d*, a removal is unavoidable. Now say the reduced level at *c* is 124.56 ft. above datum, and at *e* the lower extremity of the cross section, it is 114.99 ft. on removing the instrument to *b* to take the upper portion of cross section, of course the back sight will be at *c*, the proper reduced level of which must be *re-inserted*. Otherwise, proceeding in the ordinary way, *d* being 5.09 ft. above *c*, or 129.65 ft. above datum, will yet appear but 120.08 ft.; in consequence of the last fore sight being at *e* and not at *c*, the next back sight station. The same result would follow if only a portion, *c e*, of the cross section was taken, provided on the next remove of the instrument in continuation of main line, the back sight was taken to *c* without its reduced level being *re-inserted*. See page 72.

We have endeavoured to make ourselves understood in this matter in as few words as possible, but are yet fearful of being a little ambiguous—if so, the reader will be good enough to re-consider of what we have said, and we are sure he will not be long in discovering our meaning, which, simple though it be, demands the utmost attention of the observer when engaged in such way.

Although levelling operations appear to be most simple and elementary in their principles and mode of procedure, yet circumstances will occasionally occur, in which the ordinary mode of operation must be laid aside, modified, or other measures resorted to, according to the circumstances of the case. For instance, if Trinity high water mark was required to be transferred from the pier of a bridge during low water, it might not, perhaps, be possible to place the level staff to read above it; although at the same time it might be very possible that by *inverting* the staff, and placing the foot just up to the mark, the level of the standard could be transferred with as much ease and accuracy as at the top of the tide in the usual manner. In such a case, of course, the whole amount of the reading is to be *added* to that of the back sight, which total would then represent the height of the standard above the level of the preceding station. The same plan may be pursued in setting centre-ribs, levelling to the soffit of an arch, or other position where it may be difficult to plant the instrument above the ultimate point of level.

The subject of inverted registration of levels naturally follows the above, and which we believe, under all and every circumstance, will be found to possess decided advantages over the previously described methods.—It is rather surprising that in levelling, as in surveying, we have not accustomed ourselves to register our observations as we are working—we mean in the direction we progress. It would, indeed, be found a most inconvenient and uncertain plan if, in surveying, we were to enter our observations from the top of the page downwards; we should in such case be continually mistaking the side on which offsets were taken, and substituting the right for the left, and the contrary. But as in levelling, offsets or cross sections are not generally taken or inserted with the main line, the difficulty is not experienced until the works come to be laid out, when we are satisfied great trouble is experienced, and many errors committed, by mistaking the one side for the other; but of this we shall hereafter have occasion to speak. We now proceed to give an example of this method of registering levels, in which it will be perceived that no difference is made from the former method, with the exception of entering the observations from the bottom upwards, and which system we shall adhere to in our subsequent field-books.

In the following field-book we have introduced an example of cross sectioning, taken simultaneously, and registered with main line; a practice not to be recommended when time is of no material consequence. Where great dispatch is required, instead of taking cross sections on sidelong ground in the usual manner, it will perhaps be found preferable to take the average rate of inclination, and register it in the column of observations, as rising or falling one foot in ten, north or south of the line, or whatever it may be; observing that it will generally be found preferable to note the ground throughout either as inclination or declination; or otherwise to note it altogether on one side of the line, which will of course generally vary, at one time inclining, at another declining. The best method of procedure we consider, where time can be spared, is, to leave a stake level with ground at each point on main line where a cross section is required, and subsequently refer to them in succession when main line of section is completed. In such case our remarks on reduction of levels from an intermediate bench mark will be found useful.

Rise.	Back Sight.	Fore Sight.	Fall.	Red. Levels.	Dist. in Links	REMARKS.
2.24	3.59	1.35	....	157.86	79,90	
1.10	4.69	3.59	....	155.62	78,00	
3.32	8.01	4.69	....	154.52	77,15	Cross Section No. 10.
2.20	10.21	8.01	....	151.20	76,00	
1.39	11.60	10.21	....	149.00	75,25	
.61	12.21	11.60	....	147.61	74,65	Cross Section No. 9.
2.14	14.35	12.21	....	147.00	74,00	
1.85	2.64	.79	....	144.86	72,50	
2.48	5.12	2.64	....	143.01	71,75	Cross Section No. 8.
4.43	9.55	5.12	....	140.53	71,00	
3.08	12.63	9.55	....	136.10	70,00	
2.49	4.63	2.14	....	133.02	69,00	
2.50	7.13	4.63	....	130.53	68,20	Cross Section No. 7.
1.51	8.64	7.13	....	128.03	67,00	
1.32	9.96	8.64	....	126.52	65,15	Cross Section No. 6.
.64	10.60	9.96	....	125.20	64,00	
				124.56	63,00	Red. Level *re-inserted.*
1.02	6.53	5.51	....	129.65	300	.. *d* .. South.
2.07	8.60	6.53	....	128.63	200	...... South.
2.00	10.60	8.60	....	126.56	100	...... South.
				124.56	....	Point of Cross Section. — Cross Section. No. 5.
......	11.90	14.45	2.55	114.99	300	.. *e* .. North.
......	8.38	11.90	3.52	117.54	200	...... North.
......	4.88	8.38	3.50	121.06	100	...... North.
.26	5.14	4.88	....	124.56	63,00	Cross Section No. 5.
4.41	9.55	5.14	....	124.30	62,00	
37.97	150.25 112.28	112.28		119.89 157.86	60,00	Level and Distance of last Station.
	37.97			37.97		

Inverted Register of Levels with Cross Sections, see Plate 8.
Datum Line 100 *Feet above Low-water Spring Tides.*

What we style "retrograde levelling" is a mixture of the direct or usual mode of operation, and of the inverted process. It has occasionally occurred in our practice that, in taking a section between two points, after levelling for a distance from one extremity, we have been obliged to go to the other, and work in the contrary direction, closing on the intermediate point previously determined. In such case to arrange a simple form of registry, and reduce the levels from the original starting point, is a matter of some little moment; the choice lies in reducing the second portion of levels to an assumed bench mark, to invert the ordinary process of registration and reduction, as at page 68, or to *reverse* the registry and reduce in the usual manner. By the first method much difficulty will be experienced, and some little uncertainty in the plotting of the section and reduction to original datum without a second column of reduced levels is appended, in which the resultant levels obtained by adding or deducting the difference between the assumed and correct datum are inserted. By the second—which, perhaps, is the best—a contrary system to the ordinary process will have to be pursued throughout, and therefore liable to some objection on the score of complexity. But by the last method of *reversing* the levels, *i. e.*, entering the back sight in the fore sight column, and the contrary, all difficulty is got over when the field-registry is completed; which will then stand exactly as if the levels had been continued in the usual manner from end to end. Great care is however required, in entering the observations in the proper columns; and it should be particularly observed, that the distance and remarks are not to be entered opposite the observations, but immediately subsequent, the difference of level by the reverting process being carried thereto.

Explore levels partake of the character of check levels, inasmuch as no distances are measured or precise route taken. The object of such levelling being to ascertain the differences between any number of points; —for instance, where a road is to cross a range of hills, to know the lowest pass, and, generally speaking, to take the level of the highest and lowest ground over particular lines of country. In other cases, every point of easy recognition over a tract of country is levelled to, particularly all the roads, lanes, footpaths, &c.; and at all conspicuous points,—such as cross roads, a junction of water-courses, the angle of a wood, a summit ridge and lowest bottom of valleys, &c., or whatever points on a map can be recognised,—the reduced level from some determined datum is entered, whereby a kind of model of the country is obtained. This being obtained, a tolerably accurate profile and section of a line in any direction could be constructed, generally sufficient to admit of a decision being come to as to the most eligible; when of course a detailed survey is made and its character more strictly developed. A specimen of such a method of examining "or exploring," and subsequent profiling a country from the plan, is exhibited in figures 1 and 2, plate 9. It requires little penetration to discover that by such a document an engineer could lay out a line of road to the greatest advantage, so far as the changes in level were concerned; but the goodness of such a work would of course depend more on the geological character of the district, than the mere differences of elevation.

Flying levels are generally taken over a line of country previously determined on by the preceding method as the most advantageous. In other cases they are taken over a line supposed most favourable from a mere inspection of the country. In either case the object is to get over the ground as rapidly as possible, and if we describe its character as between trial and explore levelling we think our meaning will not be misunderstood. Chaining is of course required to give the position of roads, brooks, &c., but as long equal sights as possible are obtained on level ground. On undulating ground of course the ridges must be levelled over, but much time will be saved by *not* levelling to the bottom of the intervening valleys, and yet be able to produce a faithful document. By levelling to within about 25 or 30 feet of the bottom, and taking the next *final* fore sight across the valley and partly up the opposite ridge, getting the depth to the bottom either by estimation from the previous mean rate of descent, or by making the holder elevate the staff until within the field of view, and then adding the quantity it is so elevated. By either of these plans the level of the bottom may be ascertained within a foot or a little more, and a saving effected of two, or probably three removes of the instrument. An objection might be made to this method, that from the sights being so unequal, considerable errors would be introduced from curvature and from the instrument perhaps not being in proper adjustment. This would certainly be the case were the long sights all taken forward, or backwards; but by taking care to alternate them, equal sights would be practically obtained if an even number of similarly circumstanced valleys were crossed, or proper care taken in equalising the sights where of dissimilar character.

For the purpose of checking the results of levelling similar to the above, where of course great dispatch is required and the chances of error considerable, or to take the level of the tides into the interior of the country, much facility will be afforded by taking the level of water line from pond to pond of any still water navigation that may exist in the district. But where the traffic is great and the ponds of considerable extent, some little uncertainty will prevail as to the water line being horizontal; the same remark will apply if the principal traffic should be one way, or the wind blowing smartly in the direction of the pond's length, either of which causes will tend to keep the water line at one end above that of the other. Generally speaking, of course the water line at the upper end of a pond will be higher than at the lower end, where the supply is not lifted to the summit, and therefore a check level obtained by such means will generally show a less elevation than correct. Considerable care will be required in taking the level of the water line both below and above a lock, to be in both cases a sufficient distance from and without the *direct* influence of any recent lock of water which may have been passed. We have often availed ourselves of this economic method of checking levels, and in one particular instance remember to have carried the tide level a distance of about 50 miles in less than two days, which was found to correspond within a foot of what we had previously determined it to be. Certainly in this case the ponds were of considerable length, with very little traffic.* While on the subject of canals, we cannot help observing that it is to us a matter of great surprise that in those great lines of internal navigation constructed during the past generation, such rigid accuracy was obtained in their levels, considering the very inferior instruments in use. Were some of our lines of railway tested by a water line as the canals, (making due allowance for gradients) we fancy small credit would be allotted to the

* The levels of the various navigation ponds, and changes of railway gradients on the principal lines of communication throughout the kingdom, may be obtained from Bradshaw's Canal and Railway Map, a most useful document to every engineer or other person interested in levelling operations.

parties by whom the levels were regulated, notwithstanding the advantages possessed by them of employing such superior instruments.

On taking the profile of any line of country, of which an accurate plan is possessed, we should always recommend the sectional length to be compared with it, and in case of a difference, to be amended to correspond. Where such a test on the distances is not possessed, very great care must be exercised to ensure accuracy, and where the measurement is made over an inclined surface, the same reduced to the horizontal distance, if of that nature that the chain cannot be held level. The correction can be made either in the field or office by the proposition given at page 4, thus: "From the square of the oblique distance, subtract the square of the difference of level, and the square root of remainder will be the horizontal distance."

In the plotting of portions of a continuous section by various parties, much trouble and uncertainty will be experienced by the use of paper of various qualities, and scales of different material, some, perhaps, using box, others ivory, paper, brass, or other material—all differently affected by change of temperature. Again: some parties are much more careful than others in marking off their distances, &c., and when the separate portions of section come to be connected, a very inaccurate document is the result, requiring the utmost attention to the figured distances in the several portions of the work to prevent error. To avoid such trouble and uncertainty, *scaled* section paper * has been contrived, by which the heights and distances are marked off with certainty by every party employed, and being decimally divided both

* We believe the invention of I. K. Brunel, Esq., engineer-in-chief of the Great Western Railway; at least it was used in the preparation of some sections for that gentleman, many years since, where we first saw it employed.

longitudinally and vertically, any range of scale may be obtained. A specimen of this kind of section paper is shown at plate 9, fig. 3, with part of the section at page 68, plotted thereon. The size of the sheets generally used is imperial, and the lines printed with a very pale straw colour or blue ink; which, on plotting a section and tinting the surface line, are scarcely perceptible. By the inch square being quartered by a thicker line than the ordinary divisions, the greatest facility is afforded for adopting any decimal scale; and, on connecting the sheets, as much facility for examining a section in the whole is obtained, as, by the ordinary method; and, with this additional advantage,—that any rate of inclination may be accurately laid down without depending on the datum line being drawn or connected so as to form a right line, the paper free from buckle, or the straight edge true. We think that any engineer once using this paper for sections, would never be satisfied with any other, not only from its great convenience, but its undoubted accuracy; of course neither expansion or contraction of the paper in any way affecting the correctness of the document. Could any such plan be devised for plotting surveys, it would indeed be invaluable; but the section paper which we have been describing would, we feel no doubt, be found infinitely preferable and more accurate than plain paper, affording, as it would, a ready scale both for plotting and reference. The only difficulty about the matter would be in the transference of distances to lines running obliquely across the paper, and in the ensuring parallelism in the scale lines when many sheets came to be joined for a large plan.*

* This method of scaled paper has been usefully applied to sketch or dimension books, by which means a most accurate plan can be inserted with almost the same facility as a figured eye sketch.

With a few general remarks on the operations of levelling and plotting we shall close the present chapter. In addition to the description and in some cases sketches of bench marks appended to our registry of levels, we have found it of much advantage to number them consecutively, and frequently to insert their level on the section. Sometimes it will be found of the greatest advantage to fix on and determine the levels of bench marks *before* taking the section, particularly if the line of country should be intricate and no opposition anticipated. The greatest advantage would in such case be afforded the party in subsequently taking the section from having been over the country before, and consequently familiar with its particularities; at least much more so than he could possibly be from merely riding or walking over it, and of course by first determining the bench marks the subsequent operation of check levelling would be saved. A rough check on levels may be obtained where the line of section crosses a river more than once, by ascertaining the velocity of the stream and distance between the points of crossing, and making the necessary addition to or deduction from the reduced level at the first point of crossing. For this purpose the parties employed should invariably be directed to take the level of surface water of all rivers, streams, or canals crossed by the line of section. Where navigable rivers or canals are crossed, an absolute check may be obtained, from the level of surface water being known at all times to within a fraction of a foot, by which means we have often known errors to be discovered. The surface level of however many streams or navigations which may be crossed, whether of still or tidal water, should always be taken and noted on the section; in the latter case of course the high and low water line would be taken instead of any intermediate state of the tide, to which of course no comparative reference could be made. Levelling to trigonometrical points when lying near is one of the safest and best checks which can be obtained; for which purpose where extensive operations in levelling are to be carried on, it would be advisable to obtain the positions and levels of all such as may be within the district. We have known of surprising coincidences of level at many of the trigonometrical points both of the English and Irish Ordnance survey: and as their positions are generally such, that check levelling is most expensive and tedious, the advantages of thus using them are great. Where check levelling cannot be obtained, the greatest circumspection is requisite in making the observations, and in the subsequent computation of levels;* the same may be said of the distances, which are not generally measured with that scrupulous accuracy which is required. It is indeed of little avail that the most expensive and delicate instruments are prepared for obtaining the minutest difference of level, and used in the most skilful manner, if at the same time (as may often be observed) the measurement of the horizontal distances are entrusted to an ordinary country lout. For the natural inclination of a line of country may be altogether different from that obtained by careless chaining, however accurate the levelling may be; from this consequence we should recommend the greatest exactitude to be observed in the selection and employment of a chainman. Most engineers are exceedingly particular in

* We would strongly recommend all persons engaged in levelling operations, neither to take assistants into the field with them unless absolutely of service, or engage in conversation with any one while so occupied; or if it is impossible to avoid conversation with land occupiers or others, not to attempt to carry on operations at the same time, as, if so, serious errors will be almost certain to be committed.

obtaining an exact and careful staff-holder, but we think a good chainman to be even more essential. For some years past we have employed the same person as chainman, from his extreme exactitude in measurements, although otherwise quite ignorant; and we are satisfied engineers and surveyors would find it much to their advantage to adopt a similar plan. In surveying, the principal may (and should) follow the chain himself where he cannot place the most implicit dependence on his assistant; but for other than working sections it is next to impossible for him to do so; hence the necessity of securing the constant services of such a person as we have spoken of. In surveying we have found him of the greatest service beyond the surety of accurate chaining; for by constant observation he has become an excellent practical surveyor, although incompetent to register the measurements. After laying out the main lines of any survey, he has rarely failed in penetrating our proposed system of operations, and as readily determined on the position of false stations, and the direction of lines of detail, as we could.

Where several assistants are employed on the same section, uniformity of system should be rigidly insisted on, and the principal assistant or director of the survey, should determine bench marks, check level, and critically examine the operations of the others. Where a section is being taken in several portions simultaneously, it of course becomes necessary to reduce each to an assumed datum; without the plotting of the section should be deferred until the completion of the levels. Where the plotting is kept up with the field work, each portion is laid off from the assumed datum, and, when completed, are then either connected, or the correct general datum drawn throughout. From inaccuracy in this particular, most serious blunders have been often committed, and we doubt not the error alluded to at p. 11 arose from a similar circumstance. We remember to have been concerned in a case of this description, which arose from an intermediate portion of section having its Trinity high-water datum assumed lower than was really the case, and which by some unaccountable neglect had not been originally corrected. It was not until after many anxious examinations of the several documents that we discovered the inaccuracy, which, although then sufficiently palpable, appeared previously both alarming and difficult of explanation.

Intermediate sights in ordinary levelling need never to be taken to less than tenths of feet, it being an obvious waste of time to attempt greater minuteness in representing the surface line; but where the levels are taken to pegs driven into the ground, of course the same accuracy and minuteness must be observed as for a back or final fore sight, and may therefore be subsequently employed as bench marks. The height of the optical axis of the level when on the direct line of section may always be registered as an intermediate sight, but never as a back or final fore sight. In an intermediate sight, also, where the top of staff is below the horizontal wire, it may either be registered by estimation—if not a large quantity—or by having the staff elevated the required height above the ground line, which of course has to be added to the height of the staff. And where an intermediate sight is *above* the horizontal wire, the same may be registered as 0.00, by merely adding the *estimated* quantity above the horizontal wire, to the back and each subsequent fore sight; occasionally this plan will be found of essential service, where dispatch is requisite, and be found equally accurate as serviceable. In such case the back entry would have to be altered by the addition of the quantity above the horizontal wire, and to prevent all subsequent doubt or confusion as to the value of the entry, it will be best to cast out the difference of level at the same time.

It will of course be gathered from our previous remarks—in fact we have made the observation—that it is not necessary for the spirit level to be placed on the sectional line, but any where either to the right or left of it, as may be most convenient. In taking, therefore, the levels of any streets, it will occur at the angles that the back sight being in one street, and the fore sight in another, will not be visible, each from each ; but the difference of level between the one point and the other will nevertheless be obtained as accurately as in any other position, if they are equi-distant from the instrument. In taking a section through a town, of course such cases continually occur, and the section can only be prepared from taking the level of as many points as are accessible, and assuming the intermediate profile. In a similar manner also where landowners and occupiers are hostile, and will not permit a survey to be taken over their properties, the only method of procedure is to take the levels round without the bounds of the dissentients' property, or by the nearest road from the sectional point on one side to that on the other, and assume the profile of the intermediate space as before. None of the sights taken between the sectional points are of course to be plotted or used in any way, and may therefore be termed "accommodation sights ;" and which may often be used with advantage in taking detailed sections where the ground is precipitous or much enclosed.

In plotting sections many persons draw in a right line for their surface from station to station, thereby making an angle at each,—which is, in reality, as inaccurate as it is ill-looking; a flowing line *through all the sectional points* will approach much nearer the true profile, and take away all that harshness of appearance so frequently observed in such documents. In the plotting fractions of feet, if near to an integer, as 99.95 feet, plot to 100 feet; if 44.42, plot to 44.50 feet; by which means much time will be saved, in place of attempting to arrive at an accuracy impossible to attain, and which is not even required. The particulars of a section should be written well above the ground line, and, where possible, all on the same level, reference being made to the surface by dotted lines: many persons prefer the writing inclined at an angle of 45° to the horizon; but we think it has a much better appearance when placed vertically. The position of towns and other important places should be denoted by having the names written horizontally over the section in a conspicuous character, as also all the principal valleys and summits. By marking the hedge-rows, and such other natural and artificial boundaries as occur in the section, it will be subdivided similar to the ground plan; which will admit of the greatest advantage both for reference and many other purposes, as pointed out in the ensuing chapter. We have, indeed, prepared sections, with the reference, quantity, quality, culture, and value, inserted thereon, in addition to the ordinary information of the substrata, slopes, and cubic contents of excavations and embankments. All wood land on the sectional line should always be particularly shown, as forming a conspicuous and ready point of reference, from the memory to the ground, and from the eye to the plan.

CHAPTER V.

LEVELLING FOR CROSS SECTIONS WITH THE PLOTTING ON PLAN AND MAIN SECTION.—LEVELLING WITH THEODOLITE LONGITUDINALLY AND TRANSVERSELY, WITH FORM OF FIELD-BOOK.—LEVELLING ON MORE THAN ONE LINE.—LEVELLING THROUGH A TOWN.—DETAILED LEVELLING WITHOUT CHAINING.—LONGITUDINAL AND CROSS SECTIONING A RIVER.—SECTIO-PLANOGRAPHY.—CHOOSING A DATUM LINE, AND REDUCTION TO LEVEL OF SEA.

In the last chapter we gave a short example of levelling for cross sections simultaneously with main line, but we now proceed to give a somewhat fuller and more varied account of the operation of taking "cross sections," together with the plotting on plan and longitudinal section. Cross sections, as we before observed, may be occasionally taken simultaneously with main line,—at least so near an approximation may be made to the correct cross profile, as practically to serve all the purposes of comparison and estimation. But for actual work, where the boundary of land has to be defined to a foot, and the *prick* of the slopes to a less quantity, a subsequent operation becomes necessary, as we shall fully explain in a future chapter.

Where a cross section is required of such extent, or the ground so precipitous that it cannot be taken at one sight on each side of main line, as in the example at page 72, it is usual to leave a stump on the main section flush with ground line, the level of which is recorded at the point where the cross or transverse section is required. This point is subsequently referred to, and commonly either the reduced level of general datum, or the level of the stump taken as the datum for the cross section. In the taking of the cross section, if on both sides of the main line, our remarks at page 68, on reduction of levels from an intermediate bench mark will apply, or if on one side only, the method of procedure will be as in ordinary profiling. Examples of the first method are represented in fig. 4, pl. 9, where the cross sections are taken at right angles nearly to a tangent of the curve at the points *a* and *b* for a distance on each side of main section, the general datum being also taken for the cross profile. The plotting of the cross sections on the plan, it will be perceived, afford the greatest facility in laying out, or subsequently varying the line to suit the features of the ground. The direction of these cross sections are not exactly at right angles to the main line—of which there is no necessity—but are taken in such directions as most readily exhibit the features of the ground. We append the field-book of cross sections, but reserve that of the main line for the concluding part of our work.

Rise.	Back Sight.	Fore Sight.	Fall.	Red. Level.	Dists. in Links.	REMARKS.
....	9.60	10.80	1.20	15.41	800	
....	1.00	9.60	8.60	16.61	700	
....	1.04	11.22	10.18	25.21	600	
....	9.85	13.84	3.99	35.39	480	Sectional Point on Main Section.
....	1.33	9.85	8.52	39.38	400	
....	7.00	12.35	5.35	47.90	300	
....	.19	7.00	6.81	53.25	200	
....	......	......	......	60.06	000	
	30.01	74.66	44.65			
		30.01	......			
		44.65		44.65		

Cross Section at a, *from North to South, on the Line* c d. *Levels reduced to general Datum.*

The section of the above was taken just as in ordinary levelling, the line being first defined on the ground, passing through a point on main section, at which a stake was left. The chainage we have made continuous as in ordinary cases; the distance to which the cross section was taken on the north side being 480 links, and extended to 800 links on the south side. The reduced level of station through which the cross section was taken, was 35.39 feet above datum, which, being inserted in the proper place, enabled us to reduce the levels as at page 68. Had we not thought it necessary to reduce the levels to general datum, we should have taken it as the level of station, and reduced as in manner above; the observations in such case north of the line would all have been *plus* in the reduced column, and those south, *minus*. We give the levels of cross section at *b* reduced in this manner.

Rise.	Back Sight.	Fore Sight.	Fall.	Reduced Levels.	Dists. in Links.	REMARKS.
......	16.00	20.00	4.00	— 26.97	932	
......	9.00	16.00	7.00	— 22.97	800	
......	.37	9.00	8.63	— 15.97	700	
......	6.30	13.64	7.34	— 7.34	600	
......	6.30	10.42	4.12	38.37	B.M.	General Reduced Level.
......	3.00	6.30	3.30	0.00	500	Sectional Point on Main Section.
4.21	7.21	3.00	......	+ 3.30	400	
6.13	8.10	1.97		— .91	200	
4.60	12.70	8.10		— 7.04	60	
......	......	......	......	—11.64	00	
14.94	62.68	78.01	30.27	....		
		62.68	14.94	....		
		15.33	15.33	15.33		

Cross Section at b, *from East to West on the Line* e f. *Levels reduced to Surface Line on Main Section.*

In the above field-book we have inserted a bench mark with its reduced level to general datum, by which means the cross section might be reduced to the same, or taken independent of main section in case a level peg had not been left. The method of plotting these cross sections on plan will be perceived on reference thereto. Where the level of ground on main line at the cross sectional point is taken as the datum, the levels are plotted from the line of direction either above or below it, accordingly as they are plus or minus quantities. And where the general reduced level is taken for the datum of cross sections, the same is drawn in parallel to the line of direction, and at the proper height or depth above or below that of the cross sectional point. The same remarks apply to the plotting of the cross sections on main section as in plate 8.

Cross sections might also be taken with boning rods used in the manner pointed out at page 17; and also by taking the inclination or declination, or both, as the case might be, from the horizontal line at the required point on main section. By the latter method, it of course becomes necessary to use some instrument capable of determining altitudes, either by observing the angle and calculating the clivity therefrom, or otherwise taking it proportionally with such an instrument as described at p. 18, but which is in fact but a modification of the method by boning rods. For determining the angle the surface of any cross section forms with the horizon, many instruments may be employed, such as the quadrant, sextant, or theodolite; but the latter is by far the most preferable and accurate, although under ordinary circumstances a near approximation may be arrived at even with the pocket sextant used vertically instead of horizontally, by merely attaching a spirit level to show the horizontal line when set to zero, or otherwise by suspending a plumb-line at right angles thereto. But where extreme accuracy is requisite, and the spirit level cannot well be used, the theodolite should certainly be employed. In some cases new lines of road are proposed along the sides of a ravine where it would be scarcely possible to plant a spirit-level, and the operation of levelling with the spirit-level and staff becomes difficult and tedious in the extreme, the staff within a few feet of the instrument probably being below the line of sight. Under such circumstances the cross sections—which would be absolutely requisite—might be taken with the greatest ease and accuracy with the theodolite. But as we have not hitherto spoken of levelling with that instrument, we think it necessary to do so before proceeding further with that of cross profiling, as the directions for the levelling of the longitudinal line will go far to elucidate that of the transverse section.

In all levelling operations, where dispatch is more essential than results of extreme accuracy, or to take cross sections where the ground is precipitous, the theodolite may be used with the greatest advantage. The method of applying the theodolite to the purposes of levelling is by taking a series of vertical angles. In the absence of the spirit-level it may also be used in a similar manner to that instrument, by clamping the vertical arc at zero, setting it up at each change of level, and taking back and fore sights, exactly in the manner described for using the spirit-level. But the way in which it is principally used in levelling operations, is, by taking a series of vertical angles over the proposed line; it is however absolutely necessary, in this kind of levelling, that the instrument be adjusted to the greatest nicety, as on the extreme accuracy of its performance, coupled with great care on the part of the observer, every thing depends; the slightest negligence in observing, or carelessness in adjusting the instrument, will make the results worthless.

The line being determined on over which it is purposed taking a section, set up the theodolite at the commencement (the adjustments as explained in Chapter VI., on surveying, having been carefully attended to), and level it by means of the parallel plate screws. Then ascertain the exact height of the optical axis of the telescope, which is best done by measuring to the centre of the eye-piece; a vane staff is to be used, and the vane set carefully to such height. An assistant is then to pass along the line (not regarding the intermediate undulations of the surface) until the general inclination of the ground changes, at which spot let him set it up on a stump driven firmly into the ground. The telescope may then be moved vertically until the bisection of the vane on the staff by the cross wires is perfected; but to ensure accuracy the instrument and staff should change places, and the angle be observed as before. If there is any difference, a mean should be taken, to which set and clamp the instrument. An assistant is then to commence chaining the distance, and at each change of level in the surface of the ground,—as in common levelling operations—erect the staff, and slide up the vane until bisected by the cross wires of the telescope,—when it will be in the imaginary line connecting the instrument with the staff first erected. The assistant is to enter this distance, and the height of the vane on the staff, in his book, and continue the chainage onward until a further change in the ground requires the staff to be again erected, and so on to the end.

The most simple method of applying the data thus obtained is to prick off the observed vertical angle, which is the line of sight, and draw it in. On this line mark off the measured distances, from which let fall perpendiculars to the horizon, of the lengths denoted by the vane at the various intermediate stations; a line being traced from the extremities of such perpendiculars will represent the surface of the ground. These perpendiculars being continued until met by a horizontal line either drawn from the point at which the instrument was first set up, or any assumed datum, will show the changes of surface level throughout. But the most correct method is, to calculate the difference of level between the stations, which may be performed by the aid of a table of logarithms and logarithmic sines, in the following easy manner.—The measured distance being the hypothenuse of a right-angled triangle; therefore—*To the log. sin. of the observed angle add the logarithm of the measured distance, and their sum, deducting* 10 *from the index, will be the log. of the difference of level.* For example, the angle of elevation being 2° 40′, and on the instrument and staff changing places, the angle of depression—which it would then be—2° 42′ (the mean 2° 41′ should of course be taken), and the measured distance 7500 feet.

Log. sin.	2° 41′ . . .	8.670393
Log	7500 . . .	3.875061
	Log. of 351.118 . . .	2.545454

Difference of Level 351.1 feet.

Or the difference of level may be found thus, which is perhaps the most ready method:—*From a table of natural sines, take the sine of the angle, and multiply the hypothenuse (or measured distance) by it, which will be the difference required. The base may be also found by taking from the same table the cosine of the angle, and multiplying it by the hypothenuse.* In the above example, the distance being so great, the necessary correction for curvature and refraction should be applied; but the angle being so small, the correction might be applied to the measured distance without sensible error, the hypothenuse and base being nearly the same. The example will therefore stand thus:—

		Feet
Observed difference of level		351.118
Correction for curvature . .	1.350	
„ Refraction	.190	
		1.160
True difference of level . .		352.278

In taking a section with a theodolite, it would be desirable to take only a very few intermediate observations between the stations, in which case it would be best to calculate the bases and difference of level for each observation in the manner already described; of course adding to, or subtracting from, the calculated level, as the case might be, the difference between the ground where the instrument was planted, and each intermediate station below the line of sight, as determined by the registrations of the position of sliding vane in the assistant's book.

Field-Book for Levelling with the Theodolite.—Distances in Feet.

Angle.		INTERMEDIATES.			Hypothenuses.	Bases.	Elevation.	Depression.	Red. Level.	Observations.
Elevation.	Depression	Hypo.	Diff.	Bases.						
									100.00	Elevation of assumed Datum.
8° 30′		505	−10.0	499.5	....	....	64.64	....	164.64	
		1020	+ 2.5	1008.8	....	....	153.30	......	253.30	
		1450	−12.3	1434.1	....	....	202.00	....	302.00	
					1880	1859.36	277.88			
			Add	for curva	ture and	refracti	on. . 07			
							277.95	....	377.95	Above Datum at Station 1.
	5° 42′	600	+ 3.0	597.0	....	....	....	56.59	321.36	
		900	....	895.6	....	....	....	89.38	288.57	
		1345	−14.5	1338.4	....	....	....	148.10	229.85	
		1920	− 8.0	1910.5	....	....	....	198.70	179.25	
						2248.84				
					2260		....	224.463		
				Add	for curva	ture and	refracti	on .103		
						41.0810		224.566	153.384	Above a. at Station 2.

The plotted section of the above field-book is contained in plate 8; the scales being 330 feet to the inch horizontal, and 100 feet vertical. Columns 1 and 2 contain the angles of elevation and depression; column 6 the hypothenuses, or measured distances between the stations; column 7 the bases answering to the hypothenuses in 6; columns 3 and 5 contain the hypothenuses and bases of the intermediate stations; column 4, the difference of the intermediate stations below the line of sight;—thus at 505 feet on first measured line, the vane when bisected on the staff was at 14 feet, and the instrument being set up 4 feet above the ground, the difference was equal to 10 feet, which is entered as *minus;* at 1020 feet the vane, when bisected, was at 1.5 feet; the difference 2.5 feet is therefore entered as *plus*. In the angles of depression, these terms produce opposite effects on the quantities they are connected with to what they represent; thus at angle 5° 42′ depression at 600 feet, a rise or *plus* 3.0 feet is subtracted from the difference of level, it being a fall, which difference has to be subtracted from the previous reduced level. Columns 8 and 9 contain the elevations and depressions computed from the various stations, and the differences in column 4 are added to or subtracted from them as the case may be. Column 10 contains the reduced levels, and the remaining part of the book is left for observations. By thus calculating and reducing the levels, the section may be plotted exactly the same as if taken in the ordinary manner with the spirit-level and staff.

The above method of levelling with the theodolite, may be advantageously adopted in taking trial sections for crossing summits. One section may be carefully taken with the spirit-level, the remainder with the theodolite,—vertical angles being measured to the previously determined bench marks at the extremities; any errors that may have crept into the calculations will then be detected.

If this plan was generally adopted, how much expense would often be saved, not only in the preliminary investigations of country, but in the subsequent execution of the work. Comparative sections might be taken in every probable direction without the enormous outlay incurred by the present method, and the best in every respect undoubtedly ascertained. Such correct results would not be arrived at or expected, certainly, as with the spirit-level; but in these preliminary sections, of what consequence would be the differing a few feet from the truth, when perhaps the *crossing of a summit might be effected at a less elevation by fifty times the amount of error*. The theodolite is not here put in comparison with the spirit-level for accuracy, but for the above purposes where a near approximation is all that is necessary, it is presumed it will be found a far more advantageous instrument, more particularly in mountainous districts, where levelling by the spirit-level and staff is both tedious and costly.

Most persons in levelling with the theodolite by angles of elevation and depression, register their observations on a diagram, and afterwards insert thereon the reduction of levels and distance; thereby forming a rude kind of section, but ready at any time to be correctly plotted to scale from the figured dimensions. This method may perhaps be considered preferable to that which may appear to many to be a complicated tabular registry, but which is, in reality, as simple as the ordinary registration with spirit-level and staff. The reader will however judge for himself, and adopt that which appears to him best under the particular circumstances of his case.

Of the great superiority of the theodolite over the spirit-level for taking cross sections of precipitous ground we have already spoken, and have now therefore only to speak of the mode of operation. Where the longitudinal section is on sidelong ground of such character as spoken of at p. 89 it will be necessary to remove so far above or below the centre line as a firm footing can be obtained for the instrument, where it is to be set up as previously directed. The vane on the staff is to be set at the same height as the optical axis of the telescope, and is to be placed on the centre line at the point where the proposed cross section is required. The vertical angle and surface distances are then to be measured; and wherever the ground varies from the general inclination on either side the centre, the staff is to be set up and the height noted at which the vane is bisected. The method to be pursued in the reduction, both for level and horizontal distance, will be the same as in the previous example.

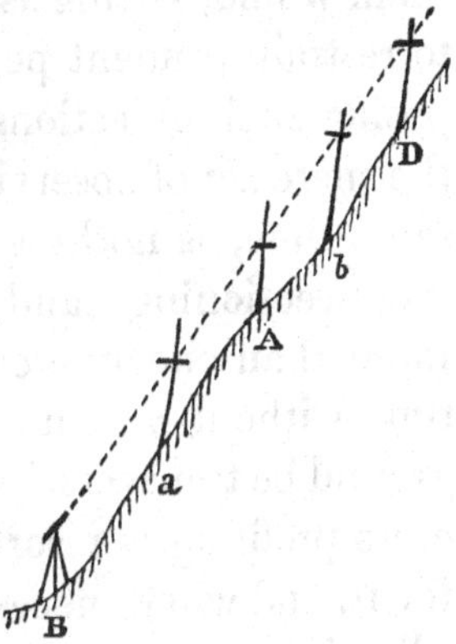

The following sketch will illustrate this method of taking cross sections. Suppose B A D, &c. to be the cross section of the ground, and *a* the centre line of the road, or sectional point on main line. At *b* below the centre line, plant the instrument, say four feet above the surface; then set up the staff at *a* with the vane at the same height, and take the vertical angle; where there is a change in the surface from the general inclination, as at *a b* D erect the staff and slide up the vane until it is bisected, and note the height. The surface distances or hypothenuses are then to be measured, when the calculations for the bases and differences of level may be made in the same manner as in the previous example. The section can be plotted to any scale as in ordinary levelling operations.

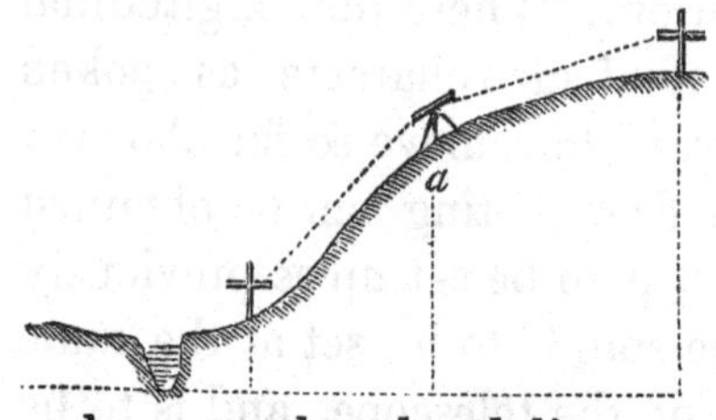

Sometimes the method shown in the accompanying diagram will be found useful where the inclination on either side the centre line is different. In such case the theodolite may be set up on the centre line, as at *a*, and the angle observed either way; the same method of reduction would be resorted to as in the preceding examples.

Levelling on two or three parallel lines at one operation has been often attempted and accomplished, but without the slightest advantage either to the operator or his employers. On the contrary, the unavoidable complexity in the observations and in the registering of them would, setting aside all other objections, be sufficient to restrain prudent persons from even a trial. But supposing such objections as we have urged to be set aside, the mere act of observing levels, on, say, three lines at the same time, is nothing more than a continuous system of cross sectioning, and with this disadvantage—that as more than one line cannot be levelled over at the same time without a removal of the instrument—except the ground be transversely level, or nearly so—it follows that cross profiling is resorted to where there is no occasion for it, and where necessary, that it is impossible to accomplish it without a distinct operation. In addition to our preceding objections, the difficulty of setting out and accurately measuring more than one line at the same time must be objectionable in the highest degree. But even when every thing which art or perseverance can do in furtherance of such a scheme; what result is obtained? why nothing more, nor so much, as might be accomplished by the aid of a few cross sections, by which parallel lines of correct section could be constructed to the full extent to which they were carried on either side the main line.

On any line of section where a river or estuary of any size or importance is crossed, it is necessary to obtain the profile of its bed, which may be arrived at by taking the level of the surface water, and sounding across it, as explained in Chapter V. of our Treatise on Surveying. Or otherwise, if the water is not too deep or strong in its current, after noting the level of high water, the staff may be pushed down to the bottom at certain intervals, and the level observed from the shore just as in ordinary levelling operations, of course making all such observations intermediate sights. Where the extent of section is not very great, a line such as is used for soundings, may be employed in determining the distances; for cross sectioning a river, it is indispensable. Where the distance is very considerable, and sounding not resorted to, the vane staff may be employed. For distances, the spirit-level may be set up at a point on one side the sectional line, determined by a measurement and meridian angle, and at each observation the bearing by compass to staff observed. By the holder planting the staff each time on the sectional line—either by observing its backward or forward direction, or otherwise by an assistant on shore placing him each time in correct position, one angle or bearing will be sufficient to determine the sectional point. To take the absolute level in continuation across a broad river or estuary, an allowance for curvature and refraction must be made on the length of sight, and great care be previously bestowed on the perfect adjusting of the instrument in all particulars. Or otherwise, where such a fore sight is taken, a back sight of equal length should be secured, or the reverse where a back sight is taken across it; by which means no correction would be required, and so much uncertainty would not exist in regard to an instrumental error. Another method is by assuming the surface water to be level at the same instant on both sides, which, if there is no wind or current at the time, will be found pretty correct.* In such case a stump should be driven exactly level with surface of water on each side at the same instant of time, on which the staff should be placed in taking both the fore and back sights. In the case of a bridge being near at hand, the level might be more readily and accurately carried round than transferred by either of the above modes; but in such case, if possible, care should be taken that it be levelled across when free from the passage of vehicles, more especially if a timber structure, as otherwise, however stable it might be, the level will be sensibly affected. Perhaps, however, the best method of transferring levels from one side of a river to the other is by reciprocal observations; for in such case what would be in excess one way would be in defect the other; consequently a mean or correct level would be the result. We may illustrate this position by the following diagram,—where *a b* is part of the earth's circumference, and *c e* the positions of a spirit-level, which are supposed to be level or equidistant from the earth's centre. Now

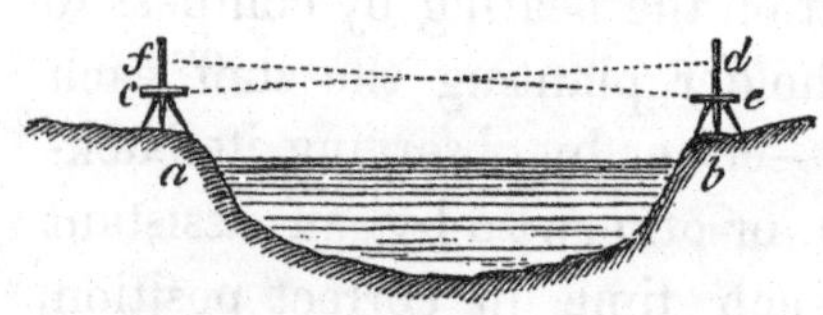

* The extreme accuracy with which levels may be tested, or transferred by the water line of a tidal river with careful observers, is surprising to those unaccustomed to such operations. We were some time since engaged in

the spirit-level being set up at both stations exactly at the same height—we will suppose 5 feet,—and that the effect of curvature on the line of sight *c d* made the reading of a staff at *b* 5.05. The spirit-level at *a* being set up at 5 feet, will therefore give a *rise* of .05 from *b* to *a*, notwithstanding we have assumed the two stations to be level. Again, the line of sight *e f* being equally affected will show a *fall* of .05 from *b* to *a*, which is contrary to the previous observation; and as both cannot be correct, —or, in other words, higher and lower at the same time, although we have assumed the observations and relative positions to be correct—we must set one error against the other, by which means the accurate or mean level will be arrived at.

In taking the longitudinal and cross sections of a river, the best plan of operation will be to level along one side the most free from obstructions, and at, say, every five chains, or oftener, leave a level stake; from which the cross section can be taken either in the manner we have pointed out at p. 97, or by such other means as the operator may think best. When the cross sections are prepared, and the levels all reduced to a datum, the longitudinal section can be drawn, either by assuming the intermediate space to be regularly inclined, or taking the level with such detail as necessary between each cross profile.

In taking the section of any road purposed to be

---

transferring Trinity high water datum from one part of the river Thames to another, and in a distance of upwards of four miles found the difference between careful levelling with the spirit-level and staff, and that determined from tidal observations after making due allowance for the invariable inclination of surface, to be but the .05 of a foot,—a coincidence as surprising as unexpected.

crossed by a railway or canal, the same method is to be pursued as in taking the cross section of any other part of the line. The plotting of such a section with the purposed manner of crossing it, is shown at plate 10.

We have before observed that the measured sectional line, where not set out with the minutest accuracy, should always be compared with the length as laid down on a good survey, and, where necessary, corrected. In such cases it will much facilitate the comparison, to note the crossings of all roads, brooks, hedges, and other natural or artificial boundaries, by which means it will be perceived whether any important deflections have been made in the line, such as would be likely to influence the section. Independent of this advantage of noting the boundaries of property, by inserting them on the section, the greatest facility is afforded for comparison of the section with the plan. For, instead of having to measure backwards or forwards from the nearest road, brook, or such other feature of the country as ordinarily inserted on sections, we have only to refer to the adjacent enclosure boundary on each, to find the exact corresponding points. By keeping this object in view we have often succeeded in taking most accurate sections in detail without chaining,—thereby avoiding in the summer season great expenses, which would otherwise have been incurred, from the damage to crops; and also the vindictive opposition which such damage, if ever so liberally paid for, seldom fails in engendering. The method which we have adopted is by considering every boundary as a determined point, and confining our divisions of distance between them, which we then either estimated by a micrometer scale, by the declination of the air bubble on the level tube scale, by taping or by pacing, as the minuteness or exigency of the work required.* Of the method of determining distances by the micrometer scale, and also by the declination of the bubble in the spirit-level tube, we shall speak fully in the next chapter, and of that by taping or pacing we need say nothing further. Respecting the registry, we enter the distance from each successive boundary to the sight or sights, as the case may be, inserting each boundary as the commencement of a new distance, and distinguishing them by a number. Opposite this number in another column the distances are inserted as taken from the plan, to which all the distances intermediate between two boundaries are added in continuity, as we have before recommended. In some cases, however, where the ground is level, or inclining or declining regularly from preceding boundary to *nearly* the next, where probably some particular feature has to be shown, it will be better to determine its position by measuring back from the next succeeding boundary and enter it as a *minus* quantity therefrom. In all cases the level on both sides of a boundary fence, or of a road or river, should be taken; more particularly if in enclosed land the ground is inclining or declining longitudinally. For by the laws of gravity the particles of surface will travel downwards, aggregate on the upper side and elevate it, while the other side will be getting depressed by the similar movement of its surface particles downwards to the next boundary. In a country which would by no means be called hilly we have never failed to observe that enclosures so situated were often many feet higher on one side the boundary than on the other, particularly if the

* A beautiful application of the telescope to the determining of distances has been recently made public by Mr. Bowman, civil engineer, of Newcastle, which promises to be of the utmost benefit to all classes of the profession. A full description of the method will be found at page 424, vol. iv. of the "Civil Engineer and Architect's Journal."

soil happened to be of a light, sandy, and drifting nature.

In the previous examples in levelling which we have given, the plan and section are detached, and totally unconnected with each other. The inconvenience this gave rise to, when numerous railway projects were recently before the legislature, was immense. Very few, except professional men, perfectly comprehend the vertical section; and still fewer, after finding their property on the plan, can refer to the corresponding part on the section,—reference and measurement having to be made by scale and compasses from some fixed point in each. Even to professional men this is a work of trouble and time, and is often of necessity inaccurately performed. Was it then probable that country gentlemen and farmers could, on inspection of the plans and sections, understand how their estates or farms would be affected? In many cases it has been known that agents of railways, presuming on their ignorance of such matters, have purposely misinformed them, for the object of gaining their acquiescence. To remedy the inconvenience, uncertainty, and fraud attendant on drawing the plan and section detached, Mr. Macneill contrived the admirable method of connecting them, termed Sectio-planography, which we proceed to describe. On reference to plate 6 there will be found in the upper part a section drawn in the usual manner, the plan appearing beneath. The strong black line on the plan represents the position of the proposed railway on the ground, and is also considered to represent a vertical section of the rail, the undulation of the ground being marked therefrom precisely in the same manner as in the section above. In fact, the line on the plan, whether straight or curved, may be considered as a datum, and after the gradient has been put on the detached section, the heights of embankment or depths of cutting—according as the rail line is above or below the surface—may be easily and accurately transferred to, or, if necessary, calculated and plotted from, such rail-line on the plan. But as the quantity or position of the various cuttings and embankments might not be readily seen on the plan, the cuttings are, or should be, coloured *red*, and the fillings *blue:* where the rails are intended to be on a level with the surface of the ground, no colour is applied. On the plans and sections which are lithographed, the trouble of colouring may be saved by representing the cuttings by vertical parallel lines, and the fillings by similar parallel horizontal lines. By this method any person, however slightly informed on such subjects, can immediately perceive on inspection of a plan, how any particular property is affected, and whether the railway passes through it in cutting, embankment, or on a level, with their depths, heights, and precise positions.

It should not, however, fail to be observed, that a distortion or inaccuracy in the surface line must necessarily occur where the line of road is curved, but which does not in any sensible degree vitiate the object in view. It never having been intended to make use of such method of laying down sections for engineering purposes, but solely to enable non-professionals to understand how their property would be affected. For all engineering uses we must come back to the detached section with the horizontal datum, which in the other case is not only curved but inclined throughout, according to the gradients, and therefore evidently unfit.

We have now a few words to say on the choosing of the datum line, which as before explained, it is necessary to fix on in the commencement of all levelling operations, as the standard to which the levels are to be referred. This level of datum or imaginary line has been taken in every possible variety by different engineers in various parts of the country, some datums having direct reference to the tides, others to mere local objects. Even where the level of the tide is taken as that of the datum, great diversity prevails; some taking the level of the highest, others only of ordinary high water spring tides; while others again take the mean level of the sea, low water spring, neap, or mean tide level. It is a difficult matter to decide on which is the best; but of this we must be certain, that where a tide level is taken for the datum, it should be unaffected by local causes, and if possible referrible to the same at all places and times. That this is not attainable by simply taking either high or low water spring or neap tide level, is certain, since at different places we find that, between points on the same coast, estuary, river, or other inlet of the sea, in one case high water mark to be on the same level, while in others, such only is the case at low water; others again at some intermediate state of the tide, and in very many instances, it never approximates. In many rivers, even the low water level at a distance from their embouchures, is above that of high water either on the coast or at their several entrances; and in pretty nearly all cases we believe it will be found that the surface at high and low water is inclined. Indeed in some cases the surface water will always be running downwards, although a difference of many feet may be effected by the rise and fall of the tide; and under such circumstances of course the surface must be inclined. Local causes undoubtedly tend to raise the level of high water very much higher at some places than at others, and which are often assisted by the action of the wind. A strong south-west or north-west wind under particular circumstances raises the tides to an extraordinary height along the south and east coast of England; and this will be the case in every position where the influence of the wind goes to assist the momentum of the tidal wave. There can be no doubt that the momentum generated by the flow of a great body of water in a particular direction,—as into a river with a funnel-shaped entrance—that the surface at the extremity of such inlet would be far above the level of that at its mouth, checked and modified, of course, by the descending current and the sinuosities and other obstructions usually existing in a river channel. The level of high water, it will therefore be granted, is not the most equable state of the tide, as the efflux cannot possibly be so affected by local or temporary causes as the influx. The level of an ordinary low water tide therefore appears to be the best suited for our purpose, —and that which was adopted throughout the English Trigonometrical Survey as being the most equable,—which may probably be hereafter modified by some formula to suit the reduction of all places to the mean level of the sea. In the interim perhaps the best plan is to endeavour to procure the mean level of the sea wherever it can be obtained free from the local influence arising from peculiar conformation of coast, or the discharge of land waters. Probably wherever it might be possible to make such observation, the mean of the oscillation of the tides would be found to be the mean level of the sea. The level of high water on shore we are certain is higher than it is at sea, or even at a short distance from the land—doubtless occasioned by the momentum of the tide wave, and the impingement of such a body of water on the shallow bottom as it approaches or sweeps along the shore. In such cases a formula promulgated by M. de la Lande, and made use of in the Trigonometrical Survey of England, for the reduction of the base to the mean level of the sea, will be found to approach pretty near to the truth; which is this:—ascertain the fall below any fixed point to the level of ordinary low water spring tides, and deduct therefrom one third of the height to which the tide rises. Thus in the Thames the difference between high and low water or rise of the tide being about 18 feet, if we first take the fall or difference of level to low water and then deduct one third of the rise, or 6 feet from the quantity, we obtain the mean level of the sea. But this will by no means be the case on rivers at a distance from the sea, or where their tides are influenced by local peculiarities; where such occur, no dependence whatever can be placed on the tide level with reference to that of the sea, or even at its mouth. Reference should therefore be made to the parties having control of the river for the elevation, it being (or should be) rarely the case where it is unknown to them. At the conclusion of our work will be found the mean rise of the tide at the principal places in the kingdom, which will perhaps be of service to many engaged in levelling operations; at all events it may perhaps serve to establish or correct the theory as laid down above, for the reduction to the level of the sea.

It may now appear an extraordinary fact, but still not the less true, that many engineers expect, or have expected, to find in their operations the level of spring tides to correspond, or nearly so, in every part of the kingdom; and that the level obtained in a river remote from the sea, but influenced by the tides, would be found to correspond when brought down to the sea. Indeed we have known of instances where sections taken from London to the coast have been unhesitatingly pronounced wrong because the level of high water spring tides, as there obtained, did not correspond with that in the Thames. What could be more fallacious than such assumption, the difference in the rise of the tide in many of these instances exceeding 20 feet.

An advantage attendant on the taking the level of low water as the datum, in addition to its being at all times more equable, is, that no *minus* quantities are ever introduced in the resultant levels; whereas, by taking the level of high water for the datum it will be of constant recurrence. Scarcely a low tract of land in the neighbourhood of the sea or of a large river but what is on a level with or below high water spring tide level, and only kept from being flooded by embankment. We might here observe that in taking the level of high water, if a conspicuous natural or artificial mark should not exist at the mean height, a good bench mark should be left in the immediate vicinity, and, where in doubt, a mean of the tides observed. Where the level of low water is taken, it of course becomes necessary to do so; as when future reference thereto was required, it might be covered by the tide, thereby occasioning inconvenience and delay.

Difficulties often arise in levelling operations from wood, water, high walls, hedges, or other features of the ground obstructing the view, or otherwise impeding progress. In the case of wood or water, the levels must either be taken round as mentioned at p. 83, or otherwise in the case of water, taken across as detailed at p. 97. The distance must of course be determined, if a correct plan on a sufficiently large scale should not be possessed by the operator. Where high walls, hedges, or other obstructions impede the view, a bench mark should be left, and the levels taken round without and beyond such. On the line of section the levels may then be carried on, by measuring the height of obstructions on arriving at them, and entering the dimensions in the level book just as though the staff had been observed in the usual manner.

CHAPTER VI.

ADVICE ON LEVELLING INSTRUMENTS.—DESCRIPTION AND ADJUSTMENTS OF THE Y LEVEL—OF TROUGHTON'S IMPROVED LEVEL—OF THE DUMPY LEVEL.—LEVELLING STAVES.—DETERMINING DISTANCES WITH THE MICROMETER SCALE, AND THE AIR BUBBLE.—CONCLUDING REMARKS.

PREVIOUS to our describing the various kinds of spirit-levels in use, we would particularly wish to impress on the minds of engineers and surveyors the necessity of possessing those of a superior quality only; inferior instruments (the correctness of the work entirely depending on the nicety of the finish) being a never-failing source of difficulty and error. Let no false economy, for the sake of saving a few pounds, induce the purchase of a common instrument, but whatever kind is used, let it be the best; then, with care, work will be always creditably and satisfactorily performed.

The staff or stand on which the level is mounted should invariably be of good substance; a tripod is the best, and made of light fir, or yellow pine, and not less than 5½ feet in height when closed. Such a stand will form a firm mounting for the instrument; the good effects of which in the field, and great advantages over the skeleton mahogany legs that are generally applied to levels (which it would be supposed were made more for show than use), will be soon apparent. Where, however, round legs are used, they should be in one length without a joint, and not tapered regularly from the top to the bottom, but swelling to an increased size at about two thirds their height, which will be found to yield greater steadiness with a diminished weight of wood.

In the operation of levelling, it will be found of service, when reversing the instrument, *to turn it only from right to left;* as it often happens (where the level is *screwed* on to the upper plate), that after observing the staff at the back station, and in attempting to turn it the reverse way to observe the forward staff, the instrument becomes partly unscrewed from the plate, and consequently its horizontality destroyed. If the staff at the back station should then have been removed, it will be necessary to go back to the last bench mark, and commence operations again; but by minding to turn it always from right to left, this never need be the case.

We commence our description of levelling instruments with that of the Y level, as being the oldest form of instrument in use, and even at the present moment very generally employed. This instrument is called *the* Y *level* from the supports which carry the telescope resembling the letter Y; but which is now being superseded by instruments of superior and more correct construction. The adjustments of this instrument are easily performed, which may account for the pertinacity with which some people assert its superiority; but on the other hand, they are also as easily deranged. The telescope is generally made to show objects erect; it is consequently darker and less distinct than those of the improved description, which have telescopes of larger diameter to show objects inverted, consequently with fewer glasses and greater brilliancy. The first adjustment in this instrument is the line of collimation; to perform which, set up the instrument in any position, open the rings which confine the telescope within the Y's, and after adjusting for distinct vision (paying no regard to the spirit-bubble), bisect some well-defined object with the cross wires and clamp the instrument firm, at the same time turning the telescope gently round as it lies on its supports or Y's, and observe if the bisection continues during a revolution of the telescope. If so, all is right; if not, alter the screws which carry the cross wires or diaphragm until the telescope will revolve in the Y's, the bisection the while remaining perfect. After adjusting for the line of collimation, carefully level the instrument by means of the parallel plate-screws; and when the spirit-bubble remains steady in the centre of the tube (the rings or clips remaining open), reverse the telescope end for end, the eye-piece being in the place previously occupied by the object-glass. If the bubble then returns to the centre of the tube, it is correct; if not, observe the end it retires to, and correct half the error by raising or lowering one end of the bubble-tube, by means of the screws by which it is attached to the telescope, and the other half by the parallel plate-screws, which will bring the bubble to the centre of the tube. Again reverse the telescope, the object-glass being in the same position as at first; the bubble should then, if correct, return to the centre of the tube; if not, alter it, as just directed, until the telescope will reverse end for end,—the bubble, in each case, returning to the centre of the tube. The next operation is to make the adjusted telescope perpendicular to the vertical axis; or, in other words, to make the instrument revolve on its stand,—the bubble remaining the while in the centre of the tube. If this should not be the case, raise or lower the milled head-screw (which carries one of the Y's, and consequently the telescope and bubble tube) one half the observed error, the other half correct as before by means of the parallel plate-screws; and if the bubble will not then remain in the centre of the tube, but retires to either end, repeat the operation until there is no perceptible difference; the instrument will then be in a proper state to observe with. But as the adjustments of the Y level are easily deranged, it is absolutely necessary to examine them frequently,—and as it is easily performed, we should recommend it *every morning;* and indeed whatever instrument is used, the surveyor will find it to his advantage to devote a few minutes every morning before proceeding to work, to examine the adjustments, and rectify, if necessary, the errors of his instrument.

A great improvement over the preceding instrument has been effected in Troughton's Improved Level, which is really an admirable instrument, and capable of taking levels to the greatest degree of accuracy; it is generally constructed to show objects inverted. Its adjustments in the hands of a beginner, or a person only accustomed to the old Y levels, may appear troublesome, tedious, and difficult; but in reality, when fully understood, they are performed with much greater facility, and with far more satisfaction to the operator. The only objection that can be made to this instrument is, that no adjustment is applied to the spirit-bubble tube; so that if, by accident or otherwise, the tube gets deranged or disturbed in the bed in which it is fixed by the maker, the line of collimation must be adapted to it, although no longer remaining in its original situation.

The adjustments necessary for the improved level are the same as those for the Y level, although from the different construction of the instrument they are differently performed. The bubble tube in this instrument has no adjustment, being fixed by the maker in the cell provided for it, which is firmly attached to the telescope. The line of collimation must therefore be adjusted to suit the bubble-tube; the most easy and correct method of doing which is, to set up the instrument on a tolerably level piece of ground, and level it by means of the parallel plate-screws. Then at a distance of three or four chains on each side of the instrument, drive a stake firmly into the ground; on these stakes alternately place the staff, and note the graduations bisected thereon by the cross hairs of the telescope. But if on reversing the telescope the air bubble does not remain in the centre of the tube, it must be brought there by the parallel plate-screws at each observation. The true level of two points will then be obtained,—the graduations bisected on the staff being equidistant from the earth's centre, however much the instrument may be out of adjustment. The instrument is then to be removed six or eight feet beyond one of the stakes, but the nearer it is placed to it, that a distinct view can be obtained, the better; again read the staffs alternately placed on the stakes, and if the readings give the same difference of level, the line of collimation will be correct. If not, raise or lower the diaphragm by the collimating screws, as the instrument looks upward or downward, until the reading on the farther staff (that on the nearest remaining the same) gives the exact difference of level of the two stakes previously ascertained. Then ascertain if the instrument will revolve on the staff head without sensible alteration in position of the bubble; if not, observe the end it retires to (as in the adjustments for the Y level), and correct half the error by the capstan-headed screws at one end of the horizontal bar, and the other half by the parallel plate-screws. If it will not then revolve without change, repeat the process, until it will turn quite round without perceptible difference; but if after repeated trials this cannot be accomplished, it must be brought into such position at each observation as pointed out above, and explained in the note, p. 64 of this volume.

Another method of collimating this instrument is by a pool of still water, in which must be driven two stakes, exactly level with the surface, and distant two or three chains from each other; then set up the instrument a few feet beyond one of the stakes, and read a staff alternately placed on each, as by the previous method, which readings, if the line of collimation is correct, will be exactly the same. If the readings are different, alter the collimating screws, until they are the same on both stakes. Or otherwise set up the instrument exactly over one of the stakes, and measure the height of the telescope above it; which should also be the reading of the staff placed on the other stake if correct. If the reading is different, alter the collimating screws, as before directed, until it is the same; then adjust the vertical axis to the plane of the instrument, as before directed. This last method is too mechanical to be entirely depended on, and should not be resorted to in preference to that previously spoken of.

The next instrument which we proceed to describe is that which, from its appearance, is well known by the name of the "Dumpy Level;" the invention of Mr. Gravatt, C. E., and has been generally commended and used by the profession; but not more so than its great merits entitle it to. This instrument is by far the most perfect for levelling operations, its short length (generally 10 inches only) rendering it less liable to accident, and its adjustments, when perfected, requiring positive violence to derange them. We have a 10-inch level of Mr. Gravatt's construction, which has not required adjusting for twelve months together, although in constant use, with the addition of having been sent per coach, at various times, many hundred miles. The large object-glass as ordinarily applied to this instrument is very advantageous,* giving greater brilliancy, and consequently distinctness, in reading the staff, and a much larger field of view; this instrument also inverts objects.

The bubble-tube, as in Troughton's improved level, is above the telescope, but, unlike it, has mechanical means of adjustment; there is also a cross bubble attached to the telescope, to assist the operator in setting up the instrument more level by means of the legs than he would otherwise be able to do; but this arrangement may of course be applied to the other descriptions of spirit-levels. The large bubble-tube is graduated to tenths of an inch, by which means the instrument can be more truly levelled, and the slightest change in the bubble instantly detected. This instrument may be adjusted in the same manner as described for Troughton's improved level, but much more correct by the method adopted by its talented inventor; thus, set it up on a tolerable level piece of ground, and, as directed for adjusting the improved level, drive in two stakes, one on each side of the instrument, at equal distances of, say, one or two chains. Read a staff placed alternately on each, when the true level of two points will be obtained, however much the spirit-level may be out of adjustment; then remove the instrument the same or double the distance beyond one of the stakes, and again set it up, and measure out the same distance beyond the instrument, as the instrument is

* Mr. Gravatt has had some levels constructed of this description with the telescope tube and object glass of a greatly increased diameter; the result has been entirely satisfactory, affording such power and brilliancy hitherto unaccomplished in instruments of such short focal length.

beyond the second stake, and drive in a third. Then read a staff placed alternately on stakes Nos. 2 and 3; and by adding or subtracting the difference of level between stakes Nos. 1 and 2, to the difference of level between stakes Nos. 2 and 3, the true level of three points will be obtained, viz., stakes 1, 2, and 3, which we will call A, B, and C, and taking stake A as the datum, suppose the difference of level to be as follows:—

	Feet.
Stake A	0.00 above Datum.
B	3.63
C	2.39

Again place the instrument a few feet beyond A, in a line with the three stakes (but the nearer the better) and carefully mark, by means of the graduations on the tube, the exact position of the bubble, so that it cannot be disturbed or the instrument altered in position without detecting it. On looking through the telescope, the staff placed on A, say, reads 4.74; on B, 0.95; and on C, 1.75. Now had the instrument been in proper adjustment when the reading at A was 4.74, the readings on B and C should have been respectively 1.11 and 2.35; the instrument, therefore, points downwards, the error at B being 0.16, and at C, 0.60. Now, was the bubble only in fault, the error at C should be three times that at B, *the distance being three times as great;* but $0.16 \times 3 = 0.48$ only; there is an error, therefore, of $0.60 - 0.48 = 0.12$, not due to the bubble. To correct this error, raise the cross-wires by means of the collimating screws, and, neglecting the actual error of level,—make the error at B only one third that at C; after a few trials, the staff at B will read 1.05 and at C, 2.17, the reading at A remaining the same. Now $1.11 - 1.05 = .06$, and $2.35 - 2.17 = 0.18$; and as $3 \times .06$ (the error at B)=0.18 (the error at C), the line of collimation will be in perfect adjustment.

What then remains to be done, is, to raise the object end of the telescope by means of the parallel plate-screws until the reading at C is 2.35, the reading at B will then be 1.11, that at A remaining as at first; then by means of the capstan-headed screws carrying the bubble tube, bring the bubble into the centre of its run. There is still another adjustment to be performed,—that of making the telescope parallel to its vertical axis, or to make it revolve on the staff-head, the bubble remaining in the centre the while; this is performed in the same manner as described for Troughton's improved level, to which the reader is referred. The operation of collimating upon levels on Mr. Gravatt's construction may appear tedious and complex, but after a few trials it will be easily understood and performed in a few minutes, but when once perfected it will scarce ever need to be repeated. But we would nevertheless recommend that the adjustments be always looked to before commencing operations, for the few minutes spent in so doing, will be amply repaid by the satisfaction produced in *knowing* that the work is correct.

In some instruments it is found very difficult to make the bubble retain a central position while the instrument is being turned round on its axis, or, as it is generally expressed, to make the instrument reverse; the operator must in this case repeatedly try to correct it, in the manner previously directed for making the telescope parallel with the vertical axis. But if after many trials the bubble will not remain in the centre of the tube, while the instrument is reversed, it must be brought to that point by means of the parallel plates at each reading of the staff.

One of the most perplexing occurrences experienced in levelling operations by the beginner, is occasioned by ebullition of the atmosphere, arising from the earth's exhalations, and the rarefaction of the atmosphere by the heat of the sun's rays. But for this there is no remedy; and when such is the case operations should if possible be suspended. But when time is an object, the effects may be palliated by taking short sights, and the mean reading *between* the limits of vibration as observed on the staff.

Instrumental parallax is another perplexing circumstance and often the cause of great errors being committed, especially in levelling operations—in fact observations of any kind, whether in surveying or levelling, are worthless if parallax exists. The causes of parallax are easily explained when the method of remedying it will immediately suggest itself. The rays of light moving in straight and parallel lines (although not actually the case, it may here be considered so without sensible error) immediately on coming in contact with the object-glass of a telescope, are bent on one side and turned from their previous straight course, converging to a point which is the focus, an image of the observed object being there formed; and for the purpose of distinguishing this object, an eye-glass of magnifying powers is applied to the telescope. To obtain a proper view of the image formed at the focus of the object-glass the focus of the eye-glass should also be at the same point; the cross-wires of the telescope appearing at the same time perfectly clear and sharp. If this is not the case *parallax* is produced; which, by looking through the telescope and bisecting any object, is at once detected by the cross-wires not remaining in contact with the object, but apparently moving with the eye, up or down, or on either side; rendering it therefore impossible to ascertain the correct bisection. To remedy this, it is necessary to move the eye-piece a very little in or out, until a clear and well-defined view of the cross wires is obtained; then turn the screw attached to the telescope (communicating motion to the slide carrying the eye-piece and cross-wires), until the view of the object is distinct. The focal point of the eye-piece will then coincide with that of the object-glass on whatever part of the optical axis it falls (the focus of the object-glass varying according to the distance of the object), and on looking through the telescope at a staff, or any well-defined object, the observer will obtain a clear and distinct view of it and the cross wires, both apparently attached, and appearing equally distant. The proof of the parallax no longer existing, will be in the moving about of the observer's eye, and no displacement taking place. But it should be observed, that the point of coincidence or focus of the glasses, varies with the state of the atmosphere, so that the adjustment which may be necessary in the morning, will often not suit for the mid-day or evening. The whole sum, substance, and correction, then, of the perplexing parallax consists in a very slight movement of the telescope eye-piece, continued until the parallax is found no longer to exist.

Changes in the atmosphere will sometimes destroy the cross wires of the diaphragm, occasioning considerable trouble and delay, and if not replaced by the operator himself, involving a total suspension of proceedings until the instrument can be transmitted to an optician and returned again. When such an occurrence takes place with ourselves, we repair it in the field or at home, as most convenient, first looseing the collimating screws and removing the diaphragm; then obtain the finest film of silk which we can draw from any garment of that material which we may have on, and pass a little gum water over it, let it dry, and gum or glue it on to the diaphragm. If we have neither glue or gum, we fix it by making a small incision on the sides of the diaphragm with a knife sufficient just to raise the metal, lay the film of silk into the notch, close it, replace the diaphragm within the telescope, and adjust for use as before.

Of levelling staves there are various descriptions; the oldest of which is that with the sliding vane, moveable by the person holding it, with a cord passing through pulleys at top and bottom, the staff being in one piece of about 12 feet in length, graduated on the face into feet and inches, or decimally into feet. There is also another kind of vane-staff similarly graduated; but instead of being in one piece, is divided into two or three sliding pieces of about five feet each, the vane in this case being moved by the hand over the first division. When the observation requires that the vane should be higher, it is effected by leaving it at the summit of the first division, and sliding that up on the second, thereby reaching 10 feet; if it should not then be high enough, the second is slided in like manner up on the third until the required height is reached. The reading on this staff is by an index on the side; thus, when the vane is at the top of the first division it will be five feet; and on being slided two feet upon the next division the reading on the index at the side of the staff will show two feet, which added to the first division of the staff will be equal to 7 feet, the height of the vane from the ground. In like manner if the second division was slided up on the third 2 feet, the height would be 12 feet; these staves are, however, now rarely used, except in remote districts where improvements have not penetrated.

The staves now in general use (the invention also of W. Gravatt, Esq. C.E.) are without any vane, the graduations, feet, tenths, and hundredths, being sufficiently distinct to enable the observer to note the reading from the instrument, and with the powerful telescopes now applied to spirit-levels, the graduations can be distinctly aud accurately noted at a distance exceeding ten chains, thereby saving much time, and obtaining more accurate results. The mechanical arrangements of Mr. Gravatt's staves are very simple; they are in three pieces, with joints similar to a fishing-rod, and when put together for use, form staves 17 feet in length, but when asunder pack conveniently for carriage. There are several modifications of Mr. Gravatt's staff, differing in their mechanical arrangements only, but all retaining the main object; that of having the graduatious sufficiently distinct to enable the observer to read off the quantities himself. The person who has most improved on Mr. Gravatt's invention is Mr. Sopwith of Newcastle, whose staff is very convenient; the graduations are nearly the same, but the decimal parts of the feet are figured; the subdivisions are also more minute. When closed it is only five feet in length, but drawing out similar to a telescope to 14 feet, a spring catch retaining each joint in its place.

A few years since we contrived a staff, which was found more convenient than any that had appeared before the public. One great fault with the improved staff is, that in reading off with an inverting telescope (which nearly all levels have now, for reasons before explained) a great liability to error exists from the figures appearing upside down; consequently, if inexperienced, careless, or in a hurry, the operator is apt to mistake one figure or division for another, thereby leading to serious errors. To remedy this inconvenience we inverted *the figures on the staff*, so that viewed through an inverting telescope they appeared in their natural order. By this simple contrivance we did away with the confusion and uncertainty previously existing, and enabled the observer instantly to note the reading with expedition and accuracy, altogether making a considerable difference both as to the quantity and quality of the work. We also applied various mechanical arrangements to this staff, differing greatly from others; the principal of which were, that being hinged, no inaccuracy could arise from the joints not being in close contact, and being throughout of the same width, the whole of the graduations were equally conspicuous, and one face folding over the other, the whole of the graduations were protected from defacement; closed for carriage it is only 5 feet in length, but opens for use to 15 feet. We also attached to the foot of this staff a universal joint connected with a flat piece of metal resting on the ground for the purpose of guarding against any change of level on the reversing of the staff at each back and fore sight. Practical men are well aware of the irremediable errors committed through the carelessness of staff-holders in this way; when the face of the staff is turned from the last forward station to become the next back, a considerable error is often occasioned through the clumsiness of the holder in pressing it into the ground, or lifting it up carelessly with clods of soil adhering to it, and again putting it down with the face reversed: the errors committed in this way are greater than are generally imagined.

The universal joint will allow the face to be turned in any direction; will adapt itself to ground of any inclination; and may be laid flat on the ground during the suspension of operations, without in the least disturbing that part resting on the surface. The staff should be pressed on the ground at each station, and on turning the face in any direction, not the slightest change in its level will take place.

Many observers attempt to remedy this source of error by putting a coin or other flat substance beneath the staff; but generally the holder is too careless or lazy to attend to it, and it is in most instances placed on the ground without any thing beneath it. Mr. Simms has contrived an iron tripod for resting the staff on, which in a great measure remedies this evil, but it is very troublesome for the man to carry.

On some staves which we have used the figures were painted sideways, and the body of the 9 filled in to prevent its being taken for a 6; this is a very good precaution so far as the prevention of error in the reading one figure for the other is concerned, but in other respects it is inefficient. To assist in the identity of the feet bisected on the staff, particularly on a dull day or towards evening when long sights are necessary, it will be found a good plan to paint the sides or border enclosing the graduations alternately white and black; by which means the feet can be readily counted up from the bottom, if the figure cannot be made out with certainty. The manner in which levelling staves are graduated is most important; as we have before observed, if carried to a considerable degree of minuteness, the divisions are wholly confused and useless at a very short distance, even with the most powerful telescopes applied to levelling instruments. Those on Sopwith's staff we consider too minute, although many have held the contrary opinion, but our readers must judge for themselves which is the best;—but of this there can be no doubt, that if at a moderate distance with an ordinary levelling telescope, the graduations are not distinctly perceptible, it is evident the minuteness of division is carried too far, and under such circumstances useless. It may not here be amiss to mention a very erroneous practice which we have observed to prevail with many persons in noting the bisection on a levelling staff: commonly the tenth of a foot is divided into ten parts by alternate spaces of white and black, thereby making hundredths of feet. Now, ordinarily, we believe, observers take the centre of each division for the hundredths instead of the top or bottom, either of which may be indifferently taken, but the same should always be adopted. The centre of the divisions in such case may be taken as thousandths, (.005) when very extreme minuteness is requisite.

There have been many forms and modifications of self reading as well as vane staves beyond what we have described; one form we remember consisted of several thin plates of metal with the alternate graduations cut through: these plates being fitted on to a rod of fir prepared by any village carpenter, and blackened on the face, made a very excellent self-registering staff; when done with, the plates were removed to be fitted to another rod when occasion required. Another form we have heard of was cylindrical, graduated all round; so that whatever position it was placed in, the graduations could be equally well seen, without the trouble and uncertainty of reversing, consequent on change of position of the spirit-level. But we think a very much better instrument than the preceding would be produced by graduating an ordinary flat staff on both sides; but in either case the greatest possible care would be requisite to ensure the parallelism of the graduations, more we think than the results would be worth.

A form of staff used by the engineers on the Grand Junction Railway we thought good on our first sight of it, but subsequently were induced to change our opinion. The feet in this description of staff are divided decimally into tenths by a horizontal line, and into hundredths by diagonals; the latter divisions being denoted by small alternate chequers of black and white in the triangular space between each two diagonal lines. The junction of two such lines being alternately .05 and .10 of a foot, the latter of which are numbered 1, 2, 3, &c. in an inverted order as on our own staff. The length is 12 or 15 feet divided by a hinge joint, and kept extended, when in use, by a slide bolt at the back; the whole staff from top to bottom is also strengthened by a wood rib at the back, which keeps it very steady and correct in windy weather.

The graduations on the face of levelling staves are ordinarily printed on paper from an engraved copperplate, and afterwards glued on to the face of the staff, and varnished as a protection from the weather. The graduations by this method are correctly set off and multiplied for the entire length of staff, which cannot be obtained perfectly when the whole has to be graduated and then painted. We have found it a good plan to paint the graduations over the paper, which is a perfect protection, and can be done by any person at a trifling trouble and expense.

Vane staves have been used with verniers to register the bisection to thousandths of feet; such were first used we believe by Mr. Bunt for tidal observations in the west of England, undertaken at the request of the British Association; subsequently improved by Mr. Allan Stevenson, and submitted to the Institution of Civil Engineers, in the session of 1841; together with his adaption of a second ball and socket to the ordinary spirit-level to prevent the necessity of working the parallel plates very obliquely, as rendered ordinarily necessary on steeply inclined ground. The additional ball and socket to the spirit-level he speaks of very confidently as an improvement, and if the instrument can be kept perfectly steady, we feel no doubt that it is a very great advantage in levelling over precipitous districts.

Another advantage this plan would effect, would be in the ability to use the spirit-level, in ranging out lines over a rugged district, which cannot now be at all managed. On level ground it may be used for such purpose as ordinarily constructed, but in some cases a pocket telescope will be found more useful. A spirit-level with an attached compass may be used as an angular instrument in a similar manner to the circumferenter, but instead of a plain needle, a floating card or ring will be found most advantageous.

The operation of determining distances by the micrometer scale is exceedingly simple, and may be described in a very few words. Attached to the eye end of the telescope, and within the diaphragm on one side, is fixed a fine comb, or scale, the parts or divisions on which subtend to a greater or less distance according as the object viewed recedes from or approaches to the observer. Now, suppose the object viewed to be a levelling staff, and if we take the whole length of it, or a portion only—say 5 feet, and note the number of divisions subtended on the micrometer-comb, or scale, at several measured distances, we obtain a comparative table by which we can at any other time determine similar distances by similar subtensions of the scale on the staff. By proportion also we can determine any other distance by noting the divisions subtended on the scale—for as the nearest number of divisions thereto in the table, is to the distance therein given, so is the number of observed divisions, to the distance required. For the purpose of determining fractional parts of the scale, which it is necessary to do in order to arrive at any thing like accuracy, a drum-headed screw with a moveable wire is used—one revolution of which moves the wire over one division of the scale; and the drumb-head of the screw being finely graduated, the fractional parts are readily determined by an index pointer.

To determine distances by the declination of the bubble, which we have often practised with complete success, it is necessary that the bubble tube be beautifully finished, its inside perfectly free from irregularities, and most accurately ground. By attention to the planting of the spirit-level, as detailed at p. 64, after levelling the instrument and reading the staff planted at a measured distance, if we decline the bubble in either direction a certain number of lines on the scale by moving one of the screws, and then note the difference of reading, we obtain a quantity which will be in proportion for any other distance. Thus, in our level, a declination of the bubble of ten lines on the scale, subtends exactly .02 on the staff at a distance of one chain; and this quantity is so exact, that the proportion remains the same for any distance at which we can read the staff.

The remainder of our treatise on Engineering Field-work consisting principally of the practical application of what we have attempted clearly to set forth in the preceding pages, we think it right to throw all together into another volume, so as to keep each department distinct, and afford the reader a ready reference to the information he needs.

CONCLUSION OF LEVELLING.

TABLES.

No. 1.

Reduction in Links and Decimals upon each Chain's Length for the following Vertical Angles.

Angle.	Reduction.	Angle.	Reduction.
3° 0′	.137	11° 45′	2.095
3 15	.161	12 0	2.185
3 30	.187	12 15	2.277
3 45	.214	12 30	2.370
4 0	.244	12 45	2.466
4 15	.275	13 0	2.553
4 30	.308	13 15	2.662
4 45	.343	13 30	2.763
5 0	.381	13 45	2.866
5 15	.420	14 0	2.970
5 30	.460	14 15	3.077
5 45	.503	14 30	3.185
6 0	.548	14 45	3.295
6 15	.594	15 0	3.407
6 30	.643	15 15	3.521
6 45	.693	15 30	3.637
7 0	.745	15 45	3.754
7 15	.800	16 0	3.874
7 30	.856	16 15	3.995
7 45	.913	16 30	4.118
8 0	.973	16 45	4.243
8 15	1.035	17 0	4.370
8 30	1.098	17 15	4.498
8 45	1.164	17 30	4.628
9 0	1.231	17 45	4.760
9 15	1.300	18 0	4.894
9 30	1.371	18 15	5.030
9 45	1.444	18 30	5.168
10 0	1.519	18 45	5.307
10 15	1.596	19 0	5.448
10 30	1.675	19 15	5.591
10 45	1.755	19 30	5.736
11 0	1.837	19 45	5.882
11 15	1.921	20 0	6.031
11 30	2.008		

No. 2.

Table of Slopes and Inclines for the following Vertical Angles.

Angle.	To one Perpendicular.	Angle.	To one Perpendicular.
0° 15′	229	8° 45′	$6\frac{1}{2}$
0 30	115	9 27	6
0 45	76	9 52	$5\frac{3}{4}$
1 0	57	10 18	$5\frac{1}{2}$
1 15	46	10 47	$5\frac{1}{4}$
1 30	39	11 19	5
1 45	33	11 53	$4\frac{3}{4}$
2 0	28	12 32	$4\frac{1}{2}$
2 15	25	13 15	$4\frac{1}{4}$
2 30	23	14 2	4
2 45	21	14 55	$3\frac{3}{4}$
3 0	19	15 56	$3\frac{1}{2}$
3 15	18	17 6	$3\frac{1}{4}$
3 38	17	18 26	3
3 35	16	19 59	$2\frac{3}{4}$
3 49	15	21 48	$2\frac{1}{2}$
4 6	14	23 58	$2\frac{1}{4}$
4 24	13	26 34	2
4 45	12	29 44	$1\frac{3}{4}$
5 0	$11\frac{1}{2}$	33 42	$1\frac{1}{2}$
5 12	11	38 40	$1\frac{1}{4}$
5 27	$10\frac{1}{2}$	45 0	1
5 42	10	53 8	$\frac{3}{4}$
6 0	$9\frac{1}{2}$	63 28	$\frac{1}{2}$
6 21	9	75 58	$\frac{1}{4}$
6 43	$8\frac{1}{2}$	78 41	$\frac{1}{5}$
7 7	8	80 32	$\frac{1}{6}$
7 36	$7\frac{1}{2}$	82 52	$\frac{1}{8}$
8 8	7	84 17	$\frac{1}{10}$

No. 3.

Difference between the Apparent and True Level for distances in Feet.

Correction in Decimals of Feet.			
Distances in Feet.	For Curvature.	For Refraction.	For Curvature and Refraction.
100	.00024	.00004	.00020
150	.00054	.00008	.00046
200	.00096	.00013	.00083
250	.00149	.00021	.00128
300	.00215	.00031	.00184
350	.00293	.00042	.00251
400	.00383	.00055	.00328
450	.00484	.00069	.00415
500	.00598	.00085	.00513
550	.00724	00103	.00621
600	.00861	.00123	.00738
650	.01010	.00144	.00866
700	.01172	.00167	.01005
750	.01345	.00192	.01153
800	.01531	.00219	.01312
850	.01728	.00247	.01481
900	.01938	.00277	.01661
950	.02159	.00308	.01851
1000	.02392	.00333	.02059
1050	.02638	.00377	.02261
1100	.02895	.00414	.02481
1150	.03164	.00452	.02712
1200	.03445	.00492	.02953
1250	.03738	.00534	.03204
1300	.04043	.00578	.03465
1350	.04361	.00623	.03738
1400	.04689	.00670	.04019
1450	.05030	.00719	.04311
1500	.05383	.00769	.04614
1550	.05748	.00821	.04927
1600	.06125	.00875	.05250
1650	.06514	.00931	.05583
1700	.06914	.00988	.05926
1750	.07327	.01047	.06280
1800	.07752	.01107	.06645
1850	.08188	.01027	.07161
1900	.08637	.01234	.07403
1950	.09098	.01300	.07798
2000	.09570	.01367	.08203

No. 4.

Difference between the Apparent and True Level for distances in Links.

Correction in Decimals of Feet.			
Distances in Links.	For Curvature.	For Refraction.	For Curvature and Refraction.
100	.00010	.00001	.00009
150	.00024	.00003	.00021
200	.00042	.00006	.00036
250	.00065	.00009	.00056
300	.00094	.00013	.00081
350	.00128	.00018	.00110
400	.00167	00024	.00143
450	.00211	.00030	.00181
500	.00261	.00037	.00224
550	.00315	.00045	.00270
600	.00375	.00054	.00321
650	.00440	.00063	.00377
700	.00511	.00073	.00438
750	.00586	.00084	.00562
800	.00667	.00095	.00572
850	.00753	.00108	.00645
900	.00844	.00121	.00723
950	.00940	.00134	.00806
1000	.01042	.00149	.00893
1050	.01149	.00164	.00985
1100	.01261	.00180	.01081
1150	.01378	.00197	.01181
1200	.01501	.00214	.01287
1250	.01628	.00233	.01395
1300	.01761	.00252	.01509
1350	.01899	.00271	.01628
1400	.02043	.00292	.01751
1450	.02191	.00313	.01878
1500	.02345	.00335	.02010
1550	.02504	.00358	.02146
1600	.02668	.00381	.02287
1650	.02837	.00405	.02432
1700	.03012	00430	.02582
1750	.03192	.00456	.02736
1800	.03377	.00482	.02895
1850	.03567	.00509	.03058
1900	.03762	.00537	.03225
1950	.03963	.00566	.03397
2000	.04169	.00596	.03573

No. 5.

Difference between the Apparent and True Level for distances in Miles.

Correction in Decimals of Feet.			
Distances in Miles.	For Curvature.	For Refraction.	For Curvature and Refraction.
¼	.0417	.0060	.0357
½	.1668	.0238	.1430
¾	.3752	.0536	.3216
1	.6670	.0953	.5717
1½	1.5008	.2144	1.2864
2	2.6680	.3811	2.2869
2½	4.1688	.5955	3.5733
3	6.0030	.8561	5.1469
3½	8.1708	1.1673	7.0035
4	10.6720	1.5246	9.1474
4½	13.5468	1.9295	11.5773
5	16.6750	2.3821	14.2929
5½	20.1769	2.8824	17.2945
6	24.0120	3.4303	20.5817
6½	28.1809	4.0258	24.1551
7	32.6830	4.6690	28.0143
7½	37.5190	5.3599	32.1591
8	42.6880	6.0997	36.5883
8½	48.1910	6.8844	41.3066
9	54.0270	7.7181	46.3089
9½	60.1971	8.5996	51.5975
10	66.7000	9.5286	57.1714
11	80.7070	11.5296	69.1774
12	96.0480	13.7211	82.3269
13	112.7230	16.1033	96.6197
14	130.7320	18.6760	112.0560
15	150.0750	21.4393	128.6357
16	170.7520	24.3931	146.3589
17	192.7630	27.5376	165.2254
18	216.1086	30.8727	185.2359
19	240.7870	34.3981	206.3889
20	266.8000	38.1143	228.6857

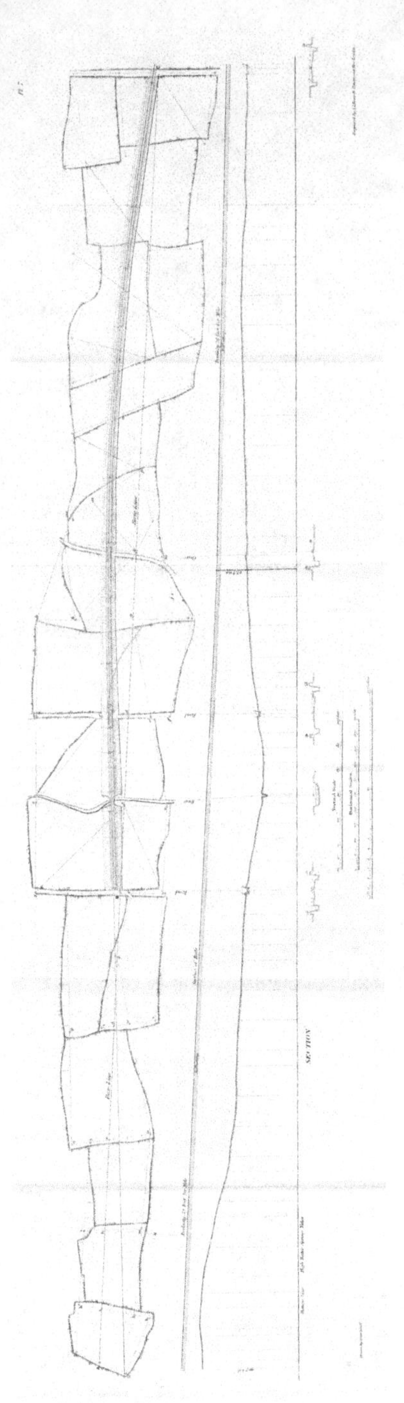

The material originally positioned here is too large for reproduction in this reissue. A PDF can be downloaded from the web address given on page iv of this book, by clicking on 'Resources Available'.

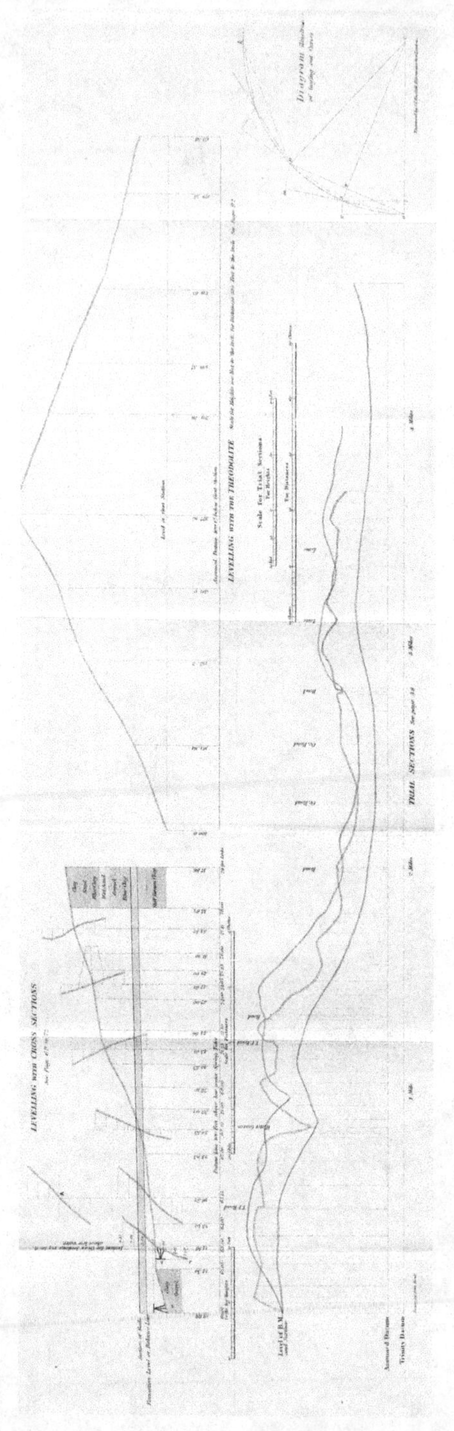

The material originally positioned here is too large for reproduction in this reissue. A PDF can be downloaded from the web address given on page iv of this book, by clicking on 'Resources Available'.

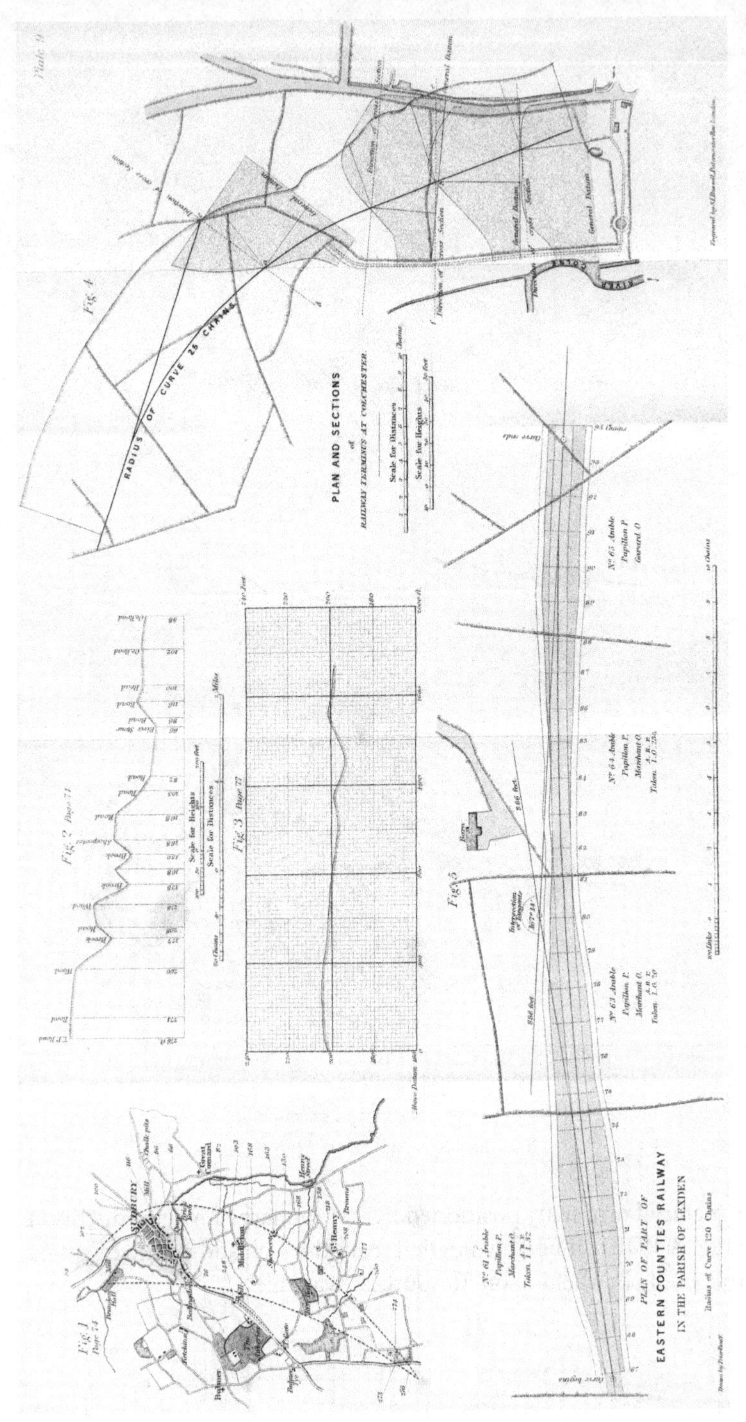

The material originally positioned here is too large for reproduction in this reissue. A PDF can be downloaded from the web address given on page iv of this book, by clicking on 'Resources Available'.

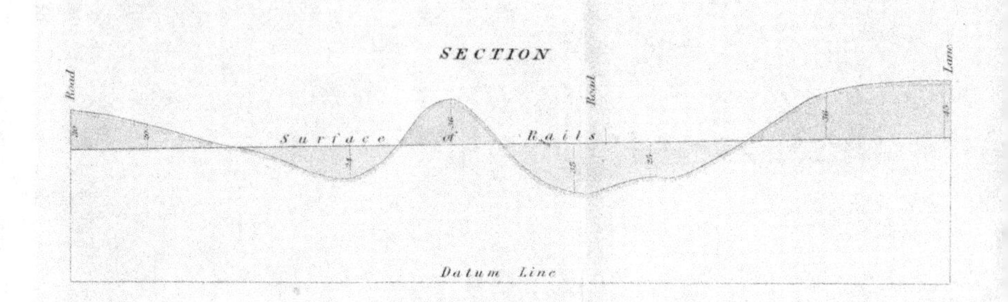

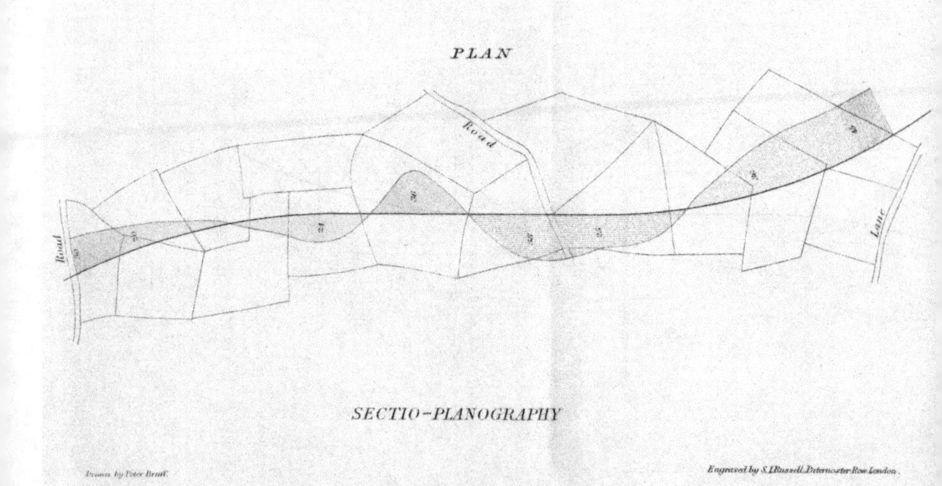

The material originally positioned here is too large for reproduction in this reissue. A PDF can be downloaded from the web address given on page iv of this book, by clicking on 'Resources Available'.

Printed in the United States
By Bookmasters